To my beloved family

Preface

Computer-aided design (CAD) tools play a vital role in modern VLSI design. Electronic simulation is widely used to verify the design and test the behavior of the circuit before fabrication. One of the major research areas in CAD is circuit simulation. Circuit simulation is to use mathematical models to predict the behavior of an electronic circuit. A circuit is usually represented by a set of partial differential equations (PDEs) or ordinary differential equations (ODEs). So, the circuit simulation actually involves solving large-scale ODEs, which sometimes takes several days or even weeks. Therefore, fast and accurate circuit simulation algorithms are needed to accelerate the simulation cycle. One way to speed up the simulation is to approximate the original system with an appropriately simplified system, which captures the main properties of the original one. This method is called model order reduction (MOR), which reduces the complexity of the original large system and generates a reduced-order model (ROM) to represent the original one. There are many existing MOR methods, but there is no method that gives the best results for all of the systems. So, each system uses the best method according to its application. So, there is still a need for novel MOR techniques. This book presents novel MOR techniques based on machine learning for linear systems. MOR is one area where "machine learning techniques" have demonstrated successful applications. We propose machine learning-based algorithms for MOR that dynamically learns the best solution. We implement our approach and compare it against state-of-the-art techniques. Our approach improves run-time without degrading the accepted accuracy.

Heliopolis, Egypt — Khaled Salah Mohamed

Contents

Chapter 1
Introduction

1 Introduction

The universe consists of matter and energy as depicted in Fig. 1.1; the aim of the engineering science is to understand both and try to model and develop any system. Many physical phenomena can be described mathematically by the same class of system. Any system can be represented by a set of continuous Partial Differential Equations (PDEs) or discrete Ordinary Differential Equations (ODEs). At the same time, any set of PDEs should be transformed into a system of ODEs which can be linear ODEs or nonlinear ODEs. So, discretization is needed which approximates the behavior of the continuous systems. For example, Maxwell's equations in electromagnetic describe the behavior of the system continuously in time and space [1]. Most CAD tools use the numerical Finite Element Method (FEM) approximation to accurately discretize in space, model and simulate these continuous structure-level VLSI systems. Solving linear ODEs results in matrix form system that can be solved using direct method such as Gaussian elimination method or indirect method (iterative methods) such as Jacobi method, and solving nonlinear ODEs can be done by Newton's method. These methods are useful for moderately sized problems.

But, solving high order (system complexity is referred to as the order of the system) or high degree of freedom of these discretized differential equations is a time-consuming process. Model order reduction (MOR) technique is a compression method to reduce the order of the full-order ODEs for fast computations and less storage requirements and at the same time it keeps the same characteristics of the full system and it should be passivity-guaranteed and stability-guaranteed model [2]. There should be a global error bound between the transfer functions of the original and reduced/compact systems. MOR is developed after the FEM discretization or any other discretization method to reduce the matrix size as illustrated in Fig. 1.2. The full-order model and the reduced model are input to a Conformance Criteria Checking step that guarantees a conformance of their characteristics as depicted in Fig. 1.3.

K. S. Mohamed, *Machine Learning for Model Order Reduction*,
https://doi.org/10.1007/978-3-319-75714-8_1

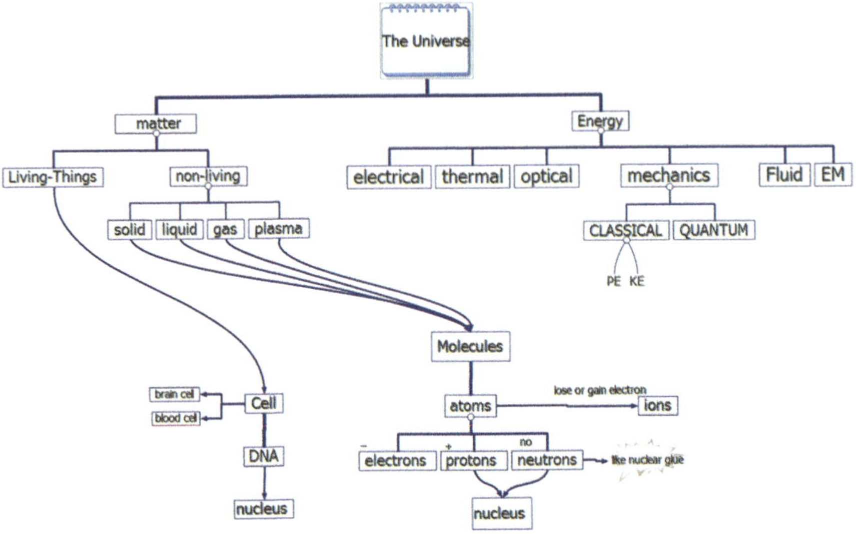

Fig. 1.1 The universe: the big picture or the high-level overview

MOR approximates the original system with an appropriately simplified system which captures the main properties of the original one. For example, the lego-based spider shown in Fig. 1.4 approximates the real spider, as it can be built in less time and acceptable accuracy.

In electromagnetic systems, is used to reduce and approximate the transfer function for single input single output (SISO) and multiple input multiple output (MIMO) systems.

The first contribution in model order reduction was published in [3, 4]. It has been discussed in [5] that a larger matrix could be simplified into a smaller one and becomes a good approximation. The basic methods in MOR were published in the 80s and 90s of the last century. Afterward, new methods were published as truncated balanced realization [6], Hankel-norm reduction method [7], and the method of proper orthogonal decomposition [8]. Asymptotic waveform evaluation [9] was the first method that is related to Krylov subspaces. PRIMA [10] is another fundamental method in MOR.

The MOR techniques are falling under four main categories as illustrated in Fig. 1.5, frequency-domain, time-domain, frequency-time domain, and optimization techniques. Frequency-domain category contains Eigen-mode analysis, moment matching, and singular value decomposition (SVD).

Eigen-mode analysis MOR has three forms: Pole-Residue Form, Pole-Zero Form, and Rational form, it is simple but computationally inefficient. Moment-matching MOR is done by matching the Taylor series coefficients (Asymptotic waveform evaluation (AWE) moment matching), but this method had a low accuracy. So, Padé approximations moment matching were later used to improve the accuracy, but due to the ill-conditioning of the matrices involved the approximate transfer function

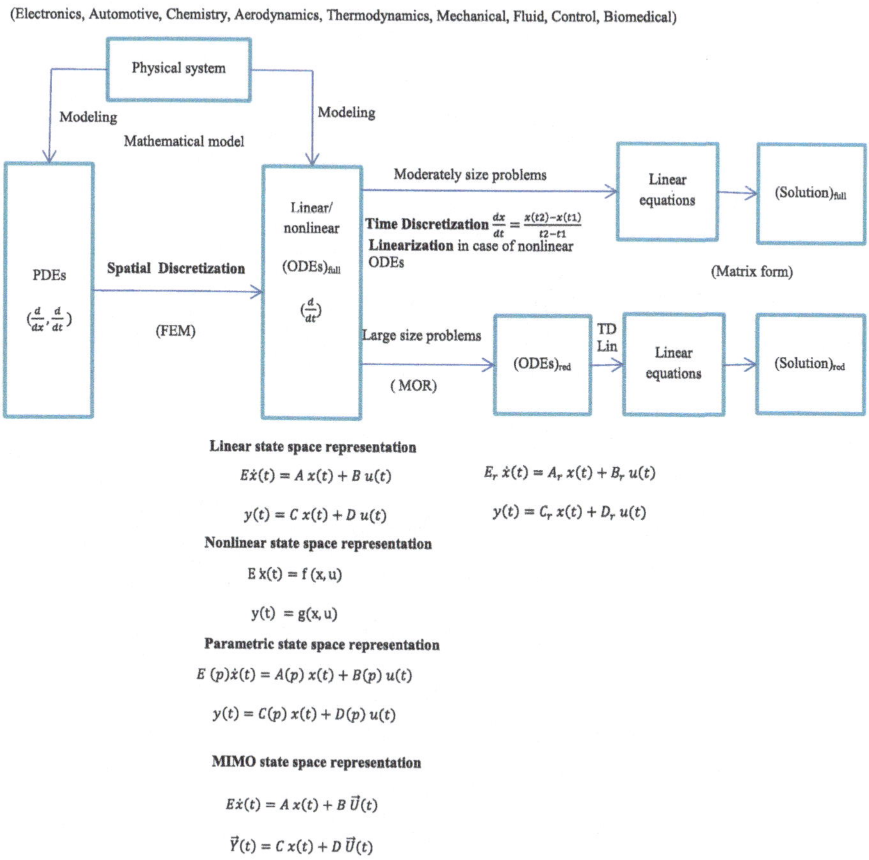

Fig. 1.2 Typical mathematical modeling flow for complex physical systems: the big picture. MOR is needed for large size problems, given an ODE of order *n*, find another ODE of order *r*, where $r << n$ with "essentially" the same "properties" and stability, passivity are preserved. $y(t)$ is the output, $u(t)$ is the input, $x(t)$ the state. We replace the set of ODEs by a smaller set of ODEs without sacrificing the accuracy of the system behavior. The universe is continuous, but computers are discrete. A computer does not understand Physics/Math equations, so we need to discretize/ sample them to be able to program them

response was found to be narrow band, leading to inaccuracies at higher orders ODEs.

Later, Krylov subspace-scheme moment matching is used to improve the ill-conditioning, it projects the problem onto a lower/smaller dimensional subspace that captures the vital aspects of the system behavior [11–13], and it is an iterative algorithm as depicted in Fig. 1.3 [14]. SVD is an approximation method based on the decomposition of the full-order system into a set of two orthogonal matrices and a diagonal matrix which contains the system's singular values. Based on the magnitude of the singular values, a direct truncation method can be applied to achieve a reduced order model. SVD is computationally inefficient [15].

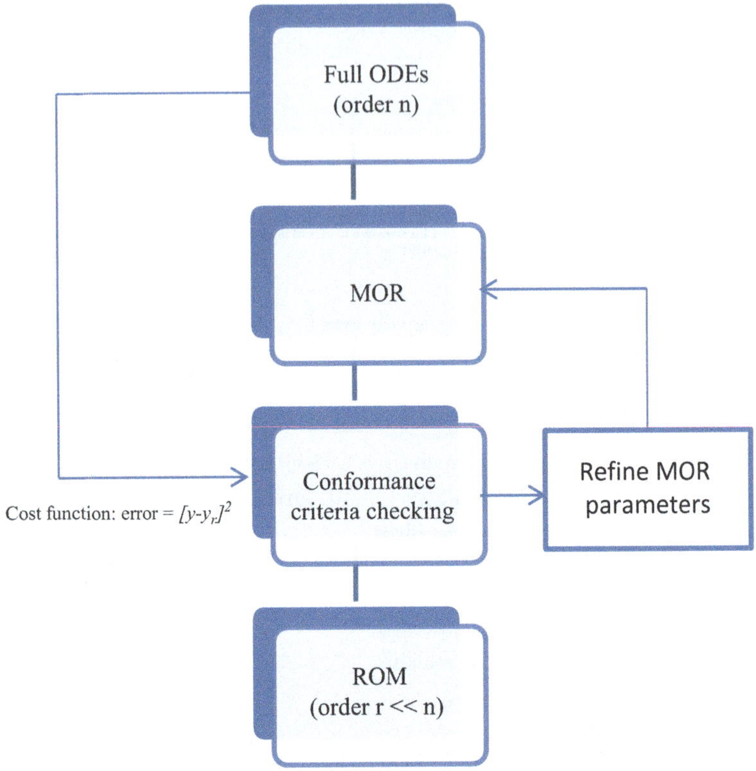

Fig. 1.3 MOR iterative method. The full-order model and the reduced model are input to a Conformance Criteria Checking step that guarantees a conformance of their characteristics

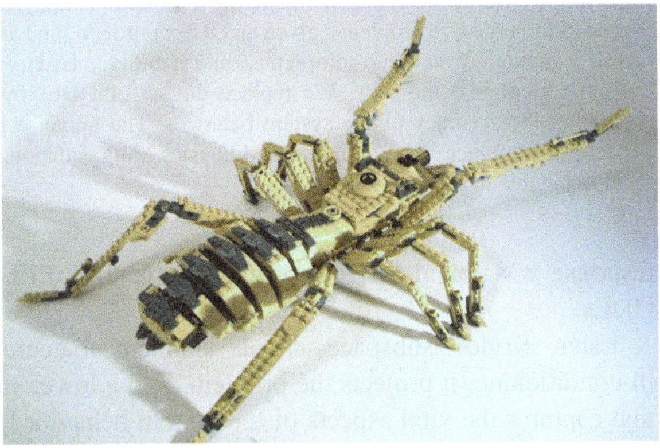

Fig. 1.4 LEGO-based spider as an example for MOR approximation

As stated earlier, MOR can be done either in frequency domain or in time domain. Frequency domain techniques such as AWE and Krylov subspace-based methods try to get a reduced order system by approximating a certain number of moments of the

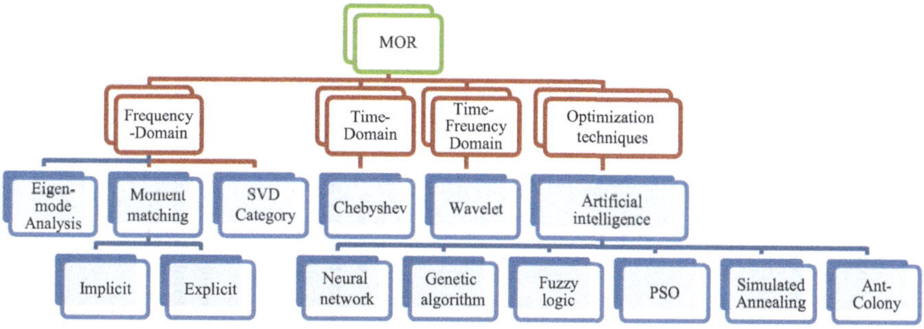

Fig. 1.5 MOR categories

original system transfer function. But, the accuracy of the output response of this reduced system in time domain cannot always be guaranteed even if the reduced transfer function can be accurate in frequency domain. So, sometime time domain MOR is used such as Chebyshev [16, 17].

Recently, wavelet-based approach is proposed for the model order reduction of linear circuits in time-frequency domain. Compared with Chebyshev reduction method, the wavelet reduction approach can achieve smaller reduced order circuits with very high accuracy [18].

The optimization techniques contain: machine learning algorithms such as neural network [19, 20], fuzzy logic, Particle swarm optimization, simulated annealing, Ant colony and genetic algorithm [21, 22].

The comparison between these different MOR methods is shown in Table 1.1. MOR represents an important research area which has been widely applied not only for electromagnetic problems, but also in other domains such as fluid dynamics, mechanics, computational biology, circuit design, control theory, biomedical applications, and so on [23]. Despite all existing MOR techniques, many unsolved problems still remain. Moreover, nonlinear MOR (NMOR), [14], and parameterized MOR (PMOR), [24], still not mature.

For NMOR, transferring these methods to the case of nonlinear systems is not straightforward. For PMOR, the high-order systems depend on parameters, for example, geometry or material parameters. Reducing the order of a large-scale system and at the same time preserving the parametric dependencies is a challenge. Moreover, there is no method that gives the best results for all of the systems. So, each system uses the best method according to its application. So, there is still a need for novel MOR techniques.

In this book, machine-learning-based algorithms such as genetic algorithm, artificial neural networks, Fuzzy logic, Particle swarm optimization, and simulated annealing have shown some promising results in solving model order reduction problem.

Table 1.1 The comparison between different MOR methods

Method		How it works	Advantage	Disadvantage
Frequency-domain				
Eigen-mode analysis	For linear systems only Pole-Residue Form: $H(s) = \sum_{i=1}^{n} \frac{R_i}{s - p_i}$ Pole-Zero Form: $H(s) = \frac{\prod_{i=1}^{n-1} (s - \zeta_i)}{\prod_{i=1}^{n} (s - p_i)}$ Rational form (point-matching): $H(s) = \frac{b_0 + b_1 s + \cdots + b_{N-1} s^{N-1}}{1 + a_1 s + \cdots + a_N s^N}$ b_i represents the numerator coefficients, a_i represents the denominator coefficients	– Drop terms with small residues – Drop terms with large poles – Find a low-order rational function matching – Take the part of the transfer function with the poles that are the closest to the imaginary axis and to throw away the others	Simple physical interpretation	Computationally inefficient
Moment-matching category	AWE (implicit) $H(s) = \sum_{i=0}^{\infty} m_i (s - s_0)^i$ With respect to $s_0 = 0$ $= m_0 + m_1 s + m_2 s^2 + \ldots$	Taylor Expansion	Simple and Computationally efficient Well conditioned	Low accuracy Does not guarantee passivity or stability
	Padé approximations $\frac{\widehat{b}_0 + \widehat{b}_1 s + \cdots + \widehat{b}_{q-1} s^{q-1}}{1 + \widehat{a}_1 s + \cdots + \widehat{a}_q s^q}$ $= m_0 + m_1 s + m_2 s^2 + \cdots + m_{2q-1} s^{2q-1}$	Rational function fitting via moment matching	Improve the accuracy for low frequency Can handle MIMO problems	Ill-conditioning. (Narrow band), leading to inaccuracies at higher orders ODEs
	Krylov subspace-scheme (explicit): (Arnoldi, PVL, PRIMA)	Projection (Change of variables)	Improve the ill-conditioning (wide-band)	Cannot handle MIMO problems Stability not preserved No Global error bound

SVD Category	Proper orthogonal decomposition (POD) method (Nonlinear system)	Decomposition	Improve the ill-conditioning	Computationally inefficient So it is used with low-order systems
	Balanced truncation (linear)	Truncation	Improve the ill-conditioning	Computationally inefficient So it is used with low-order systems
	Hankel approximation (linear)	Approximation	Improve the ill-conditioning	Computationally inefficient So it is used with low-order systems
Time-Domain				
Chebyshev		In time domain.	Preserve passivity and stability	Error increases with high order of the Chebyshev polynomial
Time-Frequency domain				
Wavelet		In time frequency domain	Computationally efficient	–
Optimization Techniques				
Machine Learning	Neural Network	This will be discussed in this book		
	Genetic algorithm			
	Fuzzy logic			
	Particle swarm optimization (PSO)			
	Simulating annealing			
	Ant-Colony			

2 Nomenclature

$\mathbb{R}$ The set of real number
$\mathbb{R}^n$ Vector of real n-tuples
$\mathbb{R}^{mxn}$ Real $m \times n$ matrices
$\rightarrow$ Input-output mapping
$\in$ Element of
$\mathcal{L}$ Laplace transform
$\mathcal{M}$ Matrix

3 Mathematical Preliminaries

The flowchart for mathematical modeling of physical systems has been shown in Fig. 1.6. A mathematical model is represented as a functional relationship of the form:

Dependent variable $= f$ (independent variables, external forces, parameters).

- *Dependent variable*: **Characteristics** that usually reflects the state of the system.
- *Independent variables*: Dimensions such as time and space along which the systems behavior is being determined.
- *External forces*: External forces influences.
- *Parameters*: Reflect the system's properties.

PDEs model the rate of change of a physical quantity with respect to both space and time variables. The order of PDE is the order of the highest derivative. Some examples of how to get the order of differential equations are shown in Table 1.2. Constants in PDEs are depending on materials and geometry. Also, not all PDEs have a solution. A differential equation is linear, if the coefficients are constant or functions only of the independent variable (time). A nonlinear differential equation is represented by nonlinear equations.

Linearization is also an important step in physical systems modeling. An example of linearization can be described in the following example. Assume that we have a system that can be described by Eq. (1.1). An equivalent linear equation can be found in Eq. (1.2).

$$z = x^2 + 4xy + 6y^2 \tag{1.1}$$

$$z = 30x + 72y - 243 \tag{1.2}$$

So, Eq. (1.1) is equivalent to Eq. (1.2) in a certain range: $8 \leq x \leq 10, 2 \leq y \leq 4$.

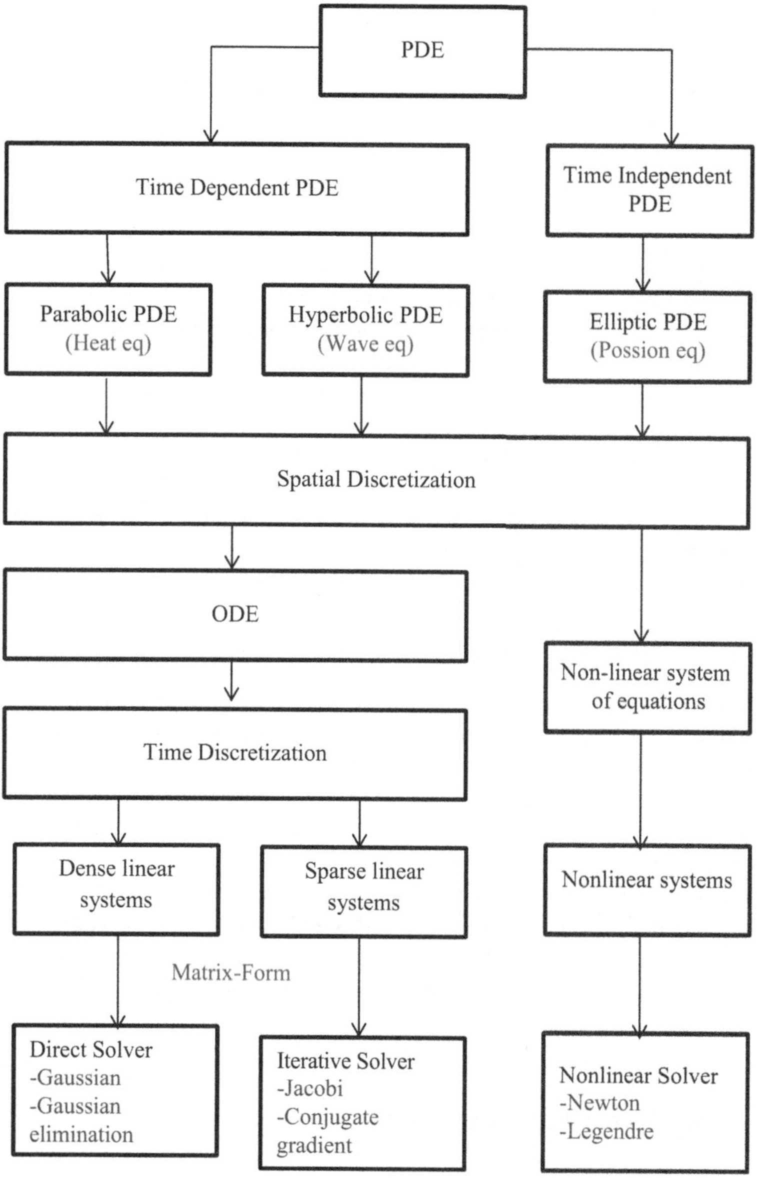

Fig. 1.6 Flow chart for modeling physical systems

4 System Theory

Any system can be described by state-space representation. The most important properties for any system are summarized in Table 1.3.

Table 1.2 Order of PDE

$\ddot{y} + (3 - t)\,\dot{y} + 6\,y = 5\,e^t$	Linear differential equation of second order
$\ddot{y} + (3 - t)\,\dot{y}^2 + 6\,y = 5\,e^t$	Nonlinear differential equation of second order

Table 1.3 Important properties of systems

Property	Description
Isotropic (scalar)	Direction independent
Homogenous	Position independent$\frac{d}{dx} = 0$
Time-invariant	Its response to the input does not change with time
static (steady-state)	Time independent $\frac{d}{dt} = 0$
Dynamic	Their behavior depends on their past evolution
Linear	Field independent $! = f(v)$ Follows the principle of superposition
Stationary	Frequency does not change with time
Causal	Future-input independent, depends only on current and past values
Deterministic	Not stochastic
Ergodic	The output is not repeated in a certain range

Moreover, all systems can be basically divided into three types:

1. Deterministic systems: These are systems for which for a given set of conditions the result can be predicted and the output does not vary much with change in initial conditions. Examples are computers.
2. Stochastic/random systems: These systems are not as reliable as deterministic systems. Their output can be predicted only for a certain range of values. Examples are genetic algorithms.
3. Chaotic systems: These systems are the most unpredictable of the three systems. Moreover they are very sensitive to initial conditions and a small change in initial conditions can bring about a great change in its output. Examples of chaotic systems are the solar system, population growth, stock market, and the weather.

4.1 State-Space Representation

The general state-space representation is in the form of:

$$E\dot{x}(t) = f(x, u) \tag{1.3}$$

$$y(t) = g(x, u) \tag{1.4}$$

where x is the unknowns states, u is the input function, y is the unknown output, $\dot{x}\,(t)$ is the rate of change, E is the system matrix.

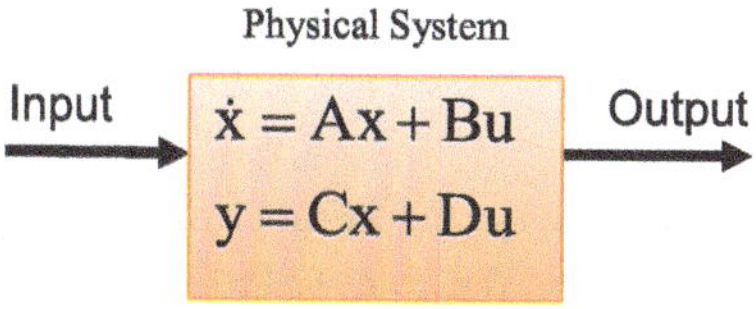

Fig. 1.7 State-space representation of a system

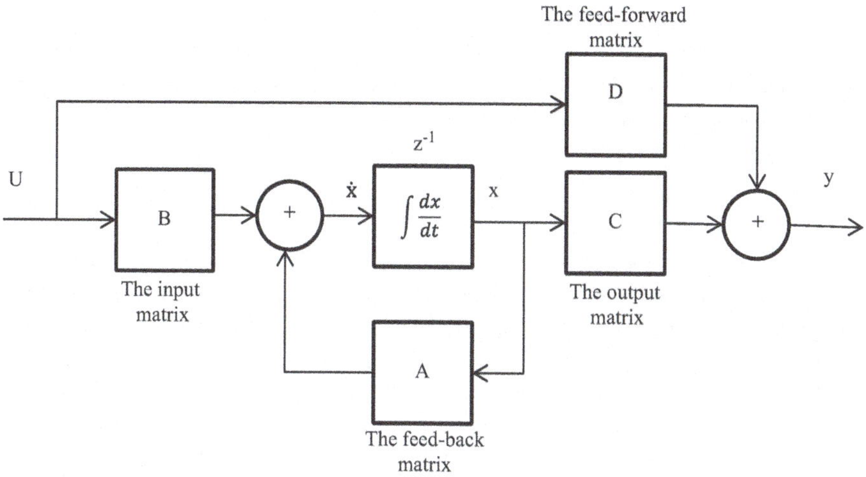

Fig. 1.8 State-space representation block diagram, where matrix A determines the order of the system

For *n*th-order linear time-invariant (LTI) systems, *f* and *g* are linear. State-space representation comes in two representations: standard state-space and modal nodal analysis (MNA) representation [25]. Standard state-space is in the [ABCD] form (Figs. 1.7 and 1.8):

$$E\dot{x}(t) = A\,x(t) + B\,u(t) \quad (1.5)$$
$$y(t) = C\,x(t) + D\,u(t) \quad (1.6)$$

where $x(t)$ is the unknowns, $u(t)$ is the input function ($u \in \mathbb{R}^m$), $y(t)$ is the unknown output ($y \in \mathbb{R}^p$), A, B, C, D, E are the system matrices. The matrix A is an $n \times n$ square matrix ($A \in \mathbb{R}^{nxn}$), B is an $n \times m$ matrix ($B \in \mathbb{R}^{nxm}$), C is a $p \times r$ matrix ($C \in \mathbb{R}^{pxr}$), D is a $p \times m$ matrix ($D \in \mathbb{R}^{pxm}$). The state differential equation relates the rate of change of the state of the system to the state of the system and the input signals as in Eq. (1.5). The outputs of a linear system is related to the state variables and the input signals by the output Eq. (1.6), where y is the set of output signals expressed in column vector form. The order of the state-space determines the complexity of the system. The matrix representation of the state-space representation is shown below:

$$\left[\frac{\dot{x}}{y}\right] = \begin{bmatrix} A & B \\ C & D \end{bmatrix} \begin{bmatrix} x \\ u \end{bmatrix} \tag{1.7}$$

Typically, the number of state variables, n, is very large so that the simulation is very slow. We want to build a ROM system, but which parts of the URM system can be discarded without changing the transfer function significantly. **Uncontrollable (not affected by the input) and unobservable (does not affect the output)** parts of the system can be removed without significantly affecting the transfer function.

The MOR methods for linear systems are mostly developed in the state-space representation, where the target system is described by state equations as below where subscript 'r' stands for reduced model:

$$\dot{x}_r(t) = A_r\, x_r(t) + B_r\, u(t) \tag{1.8}$$

$$y_r(t) = C_r\, x_r(t) + D_r\, u(t) \tag{1.9}$$

The reduction objective is to obtain A_r, B_r, C_r, D_r, x_r such that $\|y - y_r\| \leq AcceptedError$. The inputs u do not take part in the model reduction process and they are transferred from the original to the reduced model without any changes.

The state-space model of small systems may be obtained manually, but for large system an automated method is needed for state-space model generation such as Modified Nodal Analysis (MNA) in circuit simulation. Although Eqs. (1.5) and (1.6) are able to represent almost all linear dynamic systems, MNA provides easier way to represent *RLC* networks and the equations are in the form of:

$$C\dot{x}(t) + G\,x(t) = B\,u(t) \tag{1.10}$$

$$y(t) = L^T\, x(t) \tag{1.11}$$

An example of obtaining state-space model for interconnect segment is shown in Fig. 1.9.

The Laplace transform ($\mathcal{L}$) for Eqs. (1.10) and (1.11) results in [27]:

$$SX(S) = A\,X(S) + B\,U(S) \tag{1.12}$$

$$Y(S) = C\,X(S) + DU(S) \tag{1.13}$$

where $X(s)$, $Y(s)$, and $U(s)$ are the Laplace transforms of the time signals $x(t)$, $y(t)$, and $u(t)$, respectively. The transfer function H(S) which is a direct relation between the output and the input in the frequency domain is given by:

$$H(S) = \frac{Y(S)}{X(S)} = C(sE - A)^{-1}B + D \tag{1.14}$$

The Laplace transform for Eqs. (1.12) and (1.13) results in [27]:

$$H(S) = \frac{Y(S)}{X(S)} = \mathrm{L}^{\mathrm{T}}\,(G + sC)^{-1}B \tag{1.15}$$

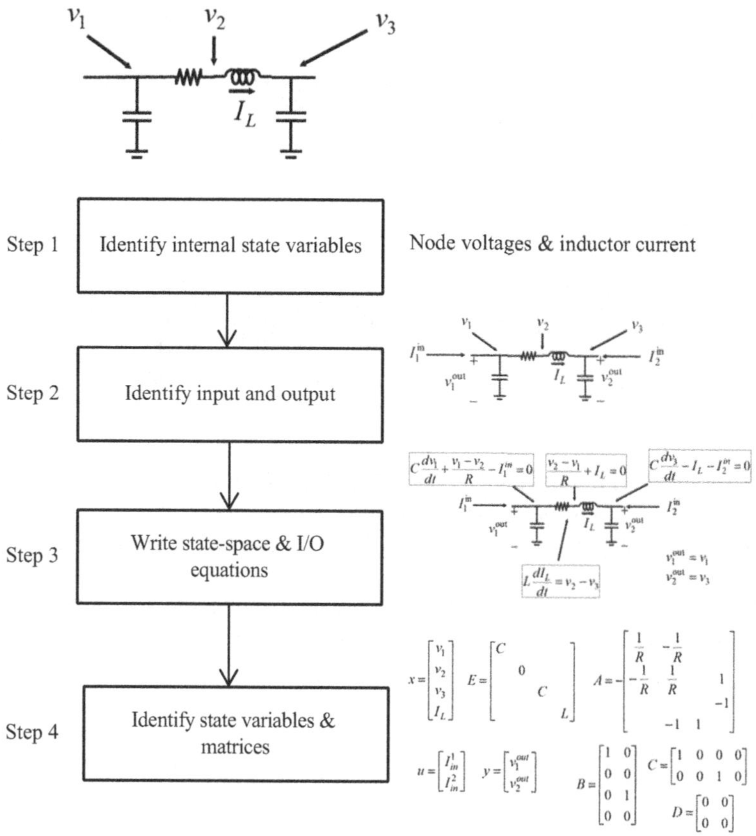

Fig. 1.9 State-space model example: Interconnect Segment [26]

For mathematical analysis, all kinds of signals should be represented in a similar form. This is done by transforming them to their Laplace form. For example, in motors the input is electrical signal whereas the output is mechanical signal. The transfer function is defined only for a linear time-invariant system, not for nonlinear system. For SISO, signals are represented in scalar form. For MIMO, signals are represented in vector form. To represent multiple inputs we expand the input $u(t)$ into a vector $U(t)$ with the desired number of inputs. We can also convert from any form to another. Converting transfer function to state-space representation or to differential equations can be described as follows:

Assume we have the following transfer function:

$$H(s) = \frac{100}{s^4 + 20s^3 + 10\,s^2 + 7s + 15} \tag{1.16}$$

The state-space representation can be deduced as follows:

- Put the denumerator in a descending order.

- Create $n \times n$ matrix which is I matrix and the last column in the coefficient of the denumerator.
- Create another $n \times 1$ matrix which is all zeros except the last row is the numerator coefficient.
- For the Y matrix: Create another $1 \times n$ matrix which is all zeros except the first row element is the numerator coefficient.

Equations (1.17) and (1.18) are the results.

$$\begin{bmatrix} \dot{x}_1 \\ \dot{x}_2 \\ \dot{x}_3 \\ \dot{x}_4 \end{bmatrix} = \begin{bmatrix} 0 & 1 & 0 & 0 \\ 0 & 0 & 1 & 0 \\ 0 & 0 & 0 & 1 \\ -15 & -7 & -10 & -20 \end{bmatrix} X + \begin{bmatrix} 0 \\ 0 \\ 0 \\ 100 \end{bmatrix} U \tag{1.17}$$

$$Y = [100 \ \ 0 \ \ 0 \ \ 0] X + [0] u \tag{1.18}$$

The differential representation can be deduced as follows:

- Put the denumerator in a descending order.
- The coefficient of the L.H.S. of the differential equation is the coefficient of the denumerator of the transfer function.
- The coefficient of the L.H.S. of the differential equation is the coefficient of the numerator of the transfer function.

Equation (1.19) is the results.

$$\frac{dx^4}{dt} + 20\frac{dx^3}{dt} + 10\frac{dx^2}{dt} + 7\frac{dx}{dt} + 15\, c = 100\, R \tag{1.19}$$

Another example can be applied on the following transfer function:

$$H(s) = \frac{s^2 + 3s + 8}{s^3 + 6s^2 + 10s + 5} = \frac{1}{s^3 + 6s^2 + 10s + 5} * s^2 + 3s + 8 \tag{1.20}$$

Equations (1.21) and (1.22) are the results.

$$\begin{bmatrix} \dot{x}_1 \\ \dot{x}_2 \\ \dot{x}_3 \end{bmatrix} = \begin{bmatrix} 0 & 1 & 0 \\ 0 & 0 & 1 \\ -5 & -10 & -6 \end{bmatrix} X + \begin{bmatrix} 0 \\ 0 \\ 1 \end{bmatrix} U \tag{1.21}$$

$$Y = [8 \ \ 3 \ \ 1] x + [0] u \tag{1.22}$$

The step response parameters are rise time, setting time, peak time, peak and overshoot. The definitions of these parameters are as below:

- Rise time: Rise time is defined as the time for the waveform to go from 0.1 to 0.9 of its final value.
- Settling time: Settling time is defined as the time for the response to reach, and stay within, 2% (or 5%) of its final value.
- Peak time: The time required to reach the first, or maximum, peak.

- Percent overshoot: The amount that the waveform overshoots the steady-state, or final, value at the peak time, expressed as a percentage of the steady-state value.

The frequency response parameters are Eigen values, damping factor, and natural frequency. The stability of a system modeled in the form of transfer function is determined by the poles of the system in the right or left hand sides of the Bode plot.

The poles are simply the denominator roots. When the model is represented using state-space approach, the Eigen values of the (A) state matrix are equivalent to the poles in the transfer function approach. Any system at a higher order than a second-order system can be modeled with a series of second- and first-order systems.

This way it is easy to find damping factor for higher order systems. Systems that have more than two poles will have a damping factor associated with each pole. The dominant pole or poles will have the associated damping factor dominate the system behavior. Natural frequency is the frequency that a second-order term would oscillate at continuously if the damping were zero.

The definitions of these parameters are as below:

- Natural Frequency: The natural frequency of a second-order system is the frequency of oscillation of the system without damping.
- Damping Ratio: The damping ratio is defined as the ratio of exponential decay frequency to natural frequency.

4.2 *Basis Functions*

We represent the unknown with basis functions. H(S) can be represented by any basis functions. For example, in Eigen-mode analysis, it is represented by rational basis function, i.e., the full-order system is expanded based on basis functions. Basis functions are mathematical functions that describe a function by decomposing it in simpler functions. Linear combination or weighted sum of these basis functions can capture any shape of functions and can model any set of data. It is like an approximation for fast computation of the important properties and features of the actual function [28]. The general form of basis functions is shown below where the function x stored as a number of coefficients (write the unknowns as summation of weighted basis functions):

$$y(x) = \sum_{n=0}^{m} a_n \Psi_n \tag{1.23}$$

where Ψ_n is a nonlinear function of x. There are an infinite number of basis functions to choose from, some of them are summarized in Table 1.4. But, the main question is how do I know what basis system to use?. The graphical representation of $y(x)$ as a function of basis functions is shown in Fig. 1.10.

Orthogonal basis functions should satisfy the following conditions:

Table 1.4 Basic basis functions

Type	Basis	Description
Polynomial	$\Psi_n = x^n$	$y(x) = a_0 + a_1x + a_2sx^2 + \ldots a_mx^m$ Known also as Taylor expansion. Useful for **non-periodic** data. There is a modified version called **splines**
Fourier	$\psi_n(t) = \begin{cases} e^{j\left(\frac{2\pi n}{T}\right)t} & 0 \le t \le T \\ 0 & t \le 0, t \ge T \end{cases}$	The Fourier basis, consisting of one and a series of pairs of sines and cosines of increasing frequency; that is, 1, sin(ωt), cos(ωt), sin($2\omega t$), cos($2\omega t$), sin($3\omega t$), cos($3\omega t$),.... The constant ω plays an important role: functions constructed with this system repeat themselves each time t increases by $2\pi/\omega$ units. Consequently, we tend to use this basis for ***periodic*** functions
Sinc	$\psi_n(t) = \sin c\left[\frac{t-n\tau}{\tau}\right] = \frac{\sin[\pi(t-n\tau)/\tau]}{\pi(t-n\tau)/\tau}$	Sinc function
Wavelet	The mother wavelet: $\psi(t) = 2\sin c(2t) - \sin c(t) = \frac{\sin 2\pi t - \sin \pi t}{\pi t}$ The child wavelet: $\psi_{a,b}(t) = \frac{1}{\sqrt{a}}\psi\left(\frac{t-b}{a}\right)$	A bit of a hybrid between the sinc and Fourier basis functions As – For **Fourier** basis functions, the number of required values defines our signal **bandwidth** B – For **sinc** basis functions, the number of required values defines our signal **timewidth** T – Wavelet basis functions allow a signal to be more **localized simultaneously** in time and frequency. The result often leads to a **fewer** number of values (i.e., fewer than Fourier or sinc) required to accurately describe the signal
Gaussian	$\psi(\mathbf{r}) = \sum_i c_i \exp\left[-\alpha_i(\mathbf{r} - \mathbf{R}_i)^2\right]$	Gaussian function
Hermite (e^x)	$H_n(x) = (-1)^n e^{x2} \frac{d}{dx^n}\left(e^{-x^2}\right)$ $H_{n+1}(x) = 2xH_n(x) - 2nH_{n-1}(x)$	Hermite function
Lagrange polynomial	$l_j(x) = \prod_{\substack{m \neq j}}^{0 \le m \ll k} \frac{x - x_m}{x_j - x_m}$	It is computationally expensive

$$\sum_{n=0}^{m} \Psi_j \Psi_k = \begin{cases} 0\, j \neq k \\ 1\, j = k \end{cases} \tag{1.24}$$

Choice of the suitable basis function is a challenge in MOR.

At steady-state ($\dot{x}(\mathrm{t}) = 0, u(t) = u = constant$), so the state-space representation yields to:

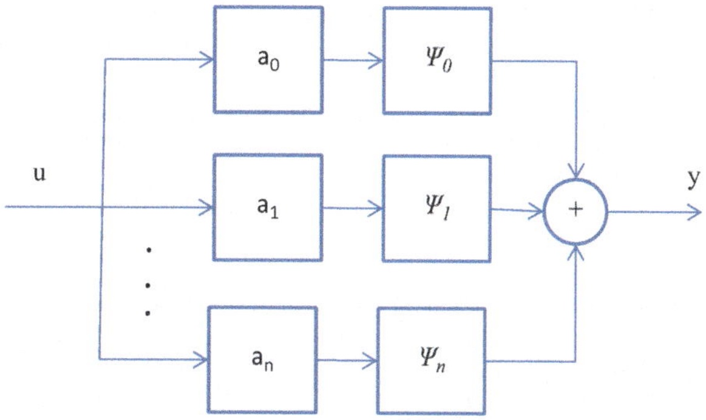

Fig. 1.10 The graphical representation of $y(x)$ as a function of basis functions. Neglect the path that have unobservable value for a_n ($a_n \approx 0$)

$$0 = Ax + Bu \tag{1.25}$$

$$y = Cx + Du \tag{1.26}$$

Equation (1.16) can be simplified to (linear system of equations):

$$Ax = -Bu = b \tag{1.27}$$

References

1. G. De Luca, G. Antonini, P. Benner, *A Parallel, Adaptive Multi-Point Model Order Reduction Algorithm*, in 22nd IEEE Conference on Electrical Performance of Electronic Packaging and Systems (EPEPS), (2013)
2. C. Wang , H. Yu , P. Li ,C. Ding , C. Sun , X. Guo , F. Zhang , Y. Zhou, Z. Yu, *Krylov Subspace Based Model Reduction Method for Transient Simulation of Active Distribution Grid* in IEEE Power and Energy Society General, Meeting (PES), PESMG, (2013)
3. J.R. Koza, *Genetic Programming On the Programming of Computers by Means of Natural Selection*, Massachusetts Institute of Technology, (1992)
4. S. Panda, J.S. Yadav, N.P. Patidar, C. Ardil, Evolutionary techniques for model order reduction of large scale linear systems. WASET Int. J. Elec. Comp. Energ. Elect. Com. Eng. **6**(9), 1105 (2012)
5. S.N. Sivanandam, S. N. Deepa, *A Comparative Study Using Genetic Algorithm and Particle Swarm Optimization for Lower Order System Modeling*, Department of Computer Science and Engineering, PSG College of Technology
6. P.O. Gutman, C.F. Mannerfelt, P. Molander, Contributions to the model reduction problem. IEEE Trans. Auto. Control **27**, 454–455 (1982)
7. E.S. Gopi, *Algorithm Collections for Digital Signal Processing Applications Using Matlab* (National Institute of Technology, Tiruchi)
8. W.H.A. Schilders, H.A. van der Vorst, J. Rommes, *Model Order Reduction: Theory, Research Aspects and Applications*, (2000)
9. P. Benner, M. Hinze, and E.J.W. ter Maten, *Model Reduction for Circuit Simulation*
10. M. Tombs, I. Postlethweite, Truncated balanced realization of a stable non-minimal state-space system. Int. J. Control **46**, 1319–1330 (1987)

11. J.E. Schutt-Ainé, P. Goh, *Comparing Fast Convolution and Model Order Reduction Methods for S-Parameter Simulation* in IEEE Electrical Design of Advanced Packaging & Systems Symposium (EDAPS), (2013)
12. B. Vernay, A. Krust, T. Maehne, G. Schröpfer, F. Pêcheux and M. M. Louërat, *A Novel Method of MEMS System-level Modeling via Multi-Domain Virtual Prototyping in SystemC-AMS*, EDAA PhD Forum, (2014)
13. Neeraj Kumar, K. J. Vinoy and S. Gopalakrishnan, *Augmented Krylov Model Order Reduction for Finite Element Approximation of Plane Wave Scattering Problems* in IEEE MTT-S International Microwave and RF Conference, IMaRC, (2013)
14. H. Aridhi, M.H. Zaki and S. Tahar, *Towards Improving Simulation of Analog Circuits using Model Order Reduction* in Design, Automation & Test in Europe Conference & Exhibition (DATE), (2012)
15. A.C. Antoulas, D.C. Sorensen, and S. Gugercin, *A Survey of Model Reduction Methods for Large-Scale Systems*, Antoulas, Sorensen, et al., (2001)
16. J.M. Wang, C.C. Chu, Q. Yu, E.S. Kuh, On projection-based algorithms for model-order reduction of interconnects. IEEE Trans. Circuit Syst. **49**(11), 1563–1585 (2002)
17. Y.L. Jiang, H.B. Chen, *Time Domain Model Order Reduction of General Orthogonal Polynomials for Linear Input-Output Systems*, in IEEE Transactions on Automatic Control, (2012)
18. Xuanzeng, L. Feng, Y. Su, W. Cai, D. Zhou, C. Chiang, *Time Domain Model Order Reduction by Wavelet Collocation Method* in Proceedings of Design, Automation and Test in Europe, (2006)
19. O.M.K. Alsmadi, Z.S. Abo-Hammour, A.M. Al-Smadi, Robust model order reduction technique for MIMO systems via ANN-LMI-based state residualization. Int. J. Circuit Theory App. **40**(4), 341–354 (2012)
20. O.M.K. Alsmadi, M.O. Abdalla, *Order Model Reduction for Two-Time-Scale Systems Based on Neural Network Estimation* in Mediterranean Conference on Control & Automation, MED '07, (2007)
21. G. Parmar, M. K. Pandey, V. Kumar, *System Order Reduction using GA for Unit Impulse Input and A Comparative Study using ISE & IRE* in International Conference on Advances in Computing, Communication and Control (ICAC3'09), (2009)
22. Z.S. Abo-Hammour, O.M.K. Alsmadi, and A.M. Al-Smadi, *Frequency-Based Model Order Reduction Via Genetic Algorithm Approach* in 7th International Workshop on Systems, Signal Processing and their Applications (WOSSPA), (2011)
23. A. Ramirez, *Reduced-Order State-Space Systems in the Dynamic Harmonic Domain* in 16th IEEE International Conference on Harmonics and Quality of Power (ICHQP), (2014)
24. M. Geuss, H. Panzer and B. Lohmann, *On Parametric Model Order Reduction by Matrix Interpolation* in European Control Conference (ECC) July 17–19, 2013, Zürich, Switzerland, (2013)
25. P. Li; H. Yu; C. Wang; C. Ding; C. Sun; Q. Zeng; B. Lei; H. Li; X. Huang, *State-space Model Generation of Distribution Networks for Model Order Reduction Application* in I EEE Power and Energy Society General Meeting (PES), (2013)
26. L. Daniel, *Model Order Reduction*, lectures notes, University of California, Berkeley, and Massachusetts Institute of Technology
27. https://www.tu-braunschweig.de/Medien-DB/numerik/benner-fassbender-mor.pdf
28. http://en.wikipedia.org/wiki/Basis_function

Chapter 2
Bio-Inspired Machine Learning Algorithm: Genetic Algorithm

1 Introduction

Genetic algorithms (GA) are the heuristic (experience-based) search and time-efficient learning and optimization techniques that mimic the process of natural evolution based on Darwinian Paradigm as depicted in Fig. 2.1. Thus genetic algorithms implement the optimization strategies by simulating evolution of species through natural selection. The nature to computer mapping is shown in Table 2.1, where each cell of a living thing contains chromosomes (strings of DNA), each chromosome contains a set of genes (blocks of DNA), and each gene determines some aspect of the organism (like eye color). In other words, parameters of the solution (genes) are concatenated to form a string (chromosome). In a chromosome, each gene controls a particular characteristic of the individual. The population evolves toward the optimal solution as depicted in Fig. 2.2. Evolution based on "survival of the fittest." Genetic algorithms are well suited for hard problems where little is known about the underlying search space. So, it is considered a robust search and optimization mechanism. The genetic algorithm used in this work consists of the following steps or operations, [1–11], and can be seen in Fig. 2.3:

1. Initialization and encoding:
   The GA starts with the creation of random strings, which represent each member in the population.
2. Evaluation (Fitness):
   The fitness used as a measure to reflect the degree of goodness of the individual is calculated for each individual in the population.
3. Selection:
   In the selection process, individuals are chosen from the current population to enter a mating pool devoted to the creation of new individuals for the next generation such that the chance of a given individual to be selected to mate is proportional to its relative fitness. This means that best individuals receive more copies in subsequent generations so that their desirable traits may be passed onto

K. S. Mohamed, *Machine Learning for Model Order Reduction*,
https://doi.org/10.1007/978-3-319-75714-8_2

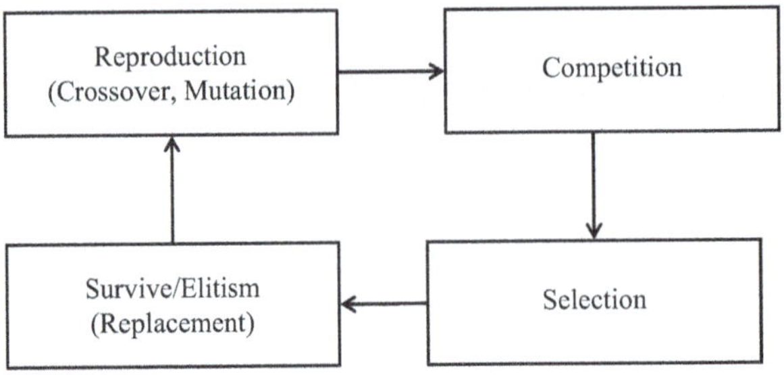

Fig. 2.1 Darwinian Paradigm**.** The procedure works as follows: Fitness Functions are the functions which are meant to be minimized, they are of particular type which satisfies a large search space. Population**:** The size of initial population, i.e., how many set of individuals are there in every generation called chromosomes. The best results from the population after competition are selected using selection functions**.** Reproduction is the main process in all evolutionary computations. Crossover takes the fraction of different chromosome from the generation and generates a new population. Mutation makes a random change in a chromosome which generates diversity. Elitism is the process through which only the best of the present population survive to form next generation

Table 2.1 The nature to computer mapping

Nature	Computer
Population	Set of solutions
Individual	Solution to a problem
Fitness	Quality of a solution
Chromosome	Encoding for a solution
Gene	Part of the encoding solution
Reproduction	Crossover

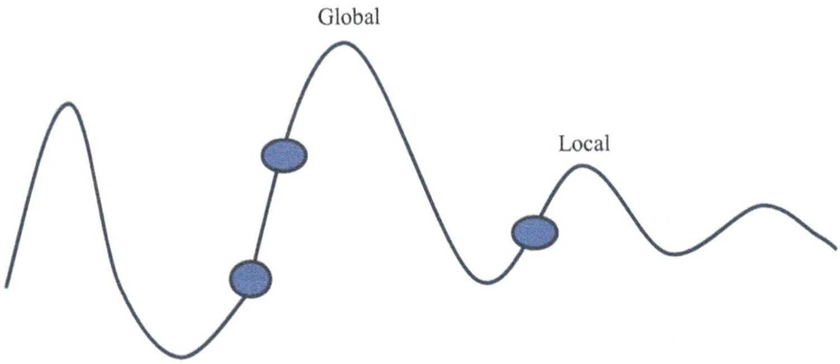

Fig. 2.2 GA searches the optimal solution in the entire search space. We chose random solutions and moves around it, until we reach global optimal not local one.

their offspring. This step ensures that the overall quality of the population increases from one generation to the next.

The different techniques used in selection process for introducing new individuals for the next generation are:

Fig. 2.3 Genetic Algorithm Chart: A GA typically operates iteratively through a simple cycle of stages: (1) creation of a population of strings, (2) evaluation of each string, (3) selection of the best strings, and (4) genetic manipulation to create a new population of strings. The fitness function is a problem-dependent. The used encoding is binary encoding

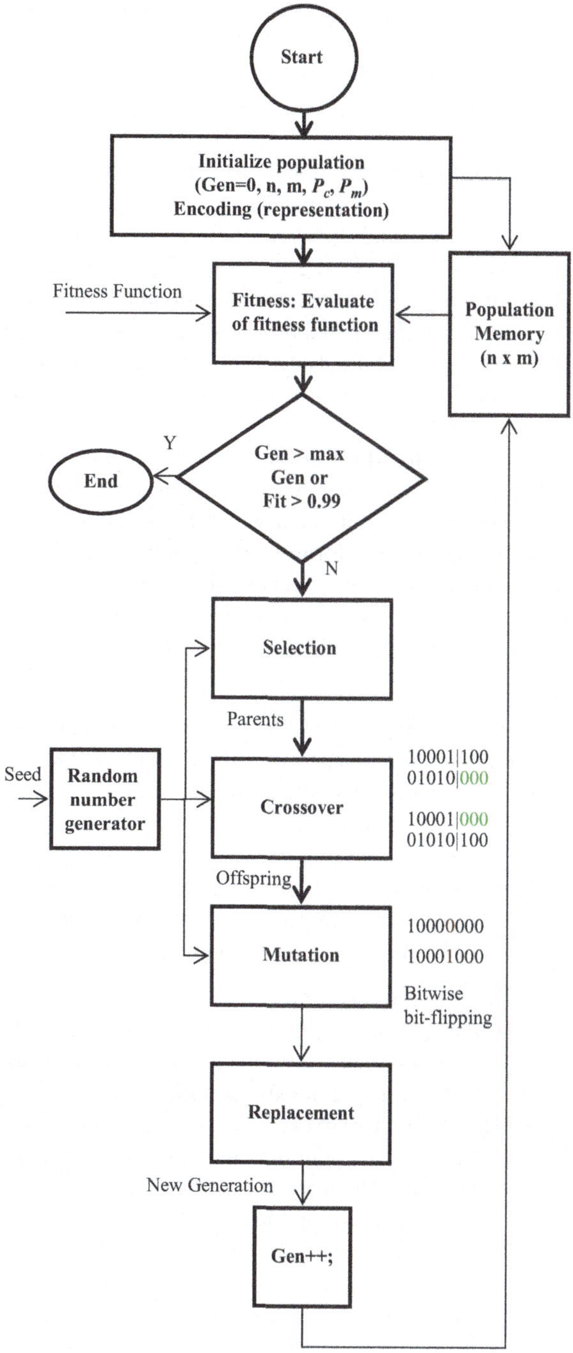

- **Elitism selection:** It is simply selecting the fitted individuals crated from the current population and passing them to the next population. This method guarantees a high probability of getting closer to the most optimum solution due to passing of the fittest chromosomes to the crossover operator producing fitter individuals.
- **Roulette Wheel Selection:** In this kind of selection the parenthood probability is directly proportional to the fitness of the individuals. Every individual is given a weight proportional to its fitness having a higher probability to be chosen in the next generation. There is a notable disadvantage in this method of selection which is if some individuals have a higher fitness than others, there is a probability that they will be chosen as parents for every individual in the next generation. The wheel rotating can be realized by generating a random number and comparing it to random selected individual in the current generation.
- **Tournament selection:** It occurs by choosing 4 or 8 or 2^n individuals from the current population and comparing each two of them randomly and so on till having one individual which is the fittest one. This process occurs as many needed time to form the new population.
- **Rank selection:** In this kind of selection all individuals in the population are ranked according to their fitness. Each individual is assigned a weight inversely proportional to the rank.

4. Crossover:

Crossover provides the means by which valuable information is shared among the population. It combines the features of two parent individuals to form two children individuals that may have new patterns compared to those of their parents and plays a central role in Gas. The crossover operator takes two chromosomes and interchanges part of their genetic information to produce two new chromosomes.

The different techniques used in crossover process for introducing new individuals for the next generation are:

- **Simple crossover**: It is a process of exchanging some genes of the chromosomes or some chromosomes of the individual depending on the representation creating new individuals of new chromosomes. It depends on crossover rate which is previously identified.
- **Heuristic crossover:** Choosing two individuals from the current individual randomly and computing the fitness of each individual and comparing them to each other. The fitter individual is passed to the next population and the other one is not passed and instead a new individual is created.
- **Arithmetic crossover**: Arithmetic crossover occurs by choosing two or more individuals randomly from the current generation and multiplying them by one random in the case of the inevitability of the presence of a certain defined domain and more than one random if there is no physical need for a defined search domain.

Table 2.2 Parameters used by the GA, the parameters are not fixed and may be changed according to the situation

Name	Symbol	Value (type)
Number of generations	Gen	200
Population size	n	50
Chromosome length	m	80 bits
Crossover probability	P_c	0.9
Mutation probability	P_m	0.01
Type of selection	–	Normal geometric, rank-based selection, roulette wheel
Type of crossover	–	Arithmetic, multi-point
Type of mutation	–	Non-uniform, flip
Termination method	–	Maximum generation, fitness > 0.99

5. Mutation:
   Mutation is often introduced to guard against premature convergence. Generally, over a period of several generations, the gene pool tends to become more and more homogeneous. The purpose of mutation is to introduce occasional perturbations to the parameters to maintain genetic diversity within the population.
6. Replacement:
   After generating the offspring's population through the application of the genetic operators to the parents' population, the parents' population is totally replaced by the offspring's population. This is known as no overlapping, generational, replacement. This completes the "life cycle" of the population.
7. Termination:
   The GA is terminated when some convergence criterion is met. Possible convergence criteria are: the fitness of the best individual so far found exceeds a threshold value; the maximum number of generations is reached. An example for the parameter used in GA is shown in Table 2.2.

2 The Proposed Technique

2.1 *Problem Formulation*

Consider we have a linear system of nth order of q inputs and r outputs described using state space in time domain as follows:

$$H = Ax(t) + Bu(t) \tag{2.1}$$

$$Y(t) = Cx(t) \tag{2.2}$$

where H is n-dimensional state vector, u is q-dimensional control vector, and Y is r-dimensional output vector with $q \leq n$ and $r \leq n$. Also, A is $n \times n$ system matrix, B is $n \times q$ input matrix, and C is $r \times n$ output matrix.

By converting the state space model from time domain to frequency domain, the transfer function can be described in Eq. (2.3):

$$G(s) = \frac{N(s)}{D(s)} = \frac{\sum_{i=0}^{n-1} A_i s^i}{\sum_{i=0}^{n} a_i s^i} \tag{2.3}$$

where $N(s)$ is the numerator and $D(s)$ is the denominator of the higher order system. Also, A_i and a_i are the coefficients of numerator and denominator, respectively.

The problem is to find a mth lower order model $R^m(s)$, where $m < n$ in the following form:

$$R(s) = \frac{N^m(s)}{D^m(s)} = \frac{\sum_{i=0}^{n-1} B_i s^i}{\sum_{i=0}^{n} b_i s^i} \tag{2.4}$$

In addition, we must take into consideration that the reduced model maintains the characteristics of the original transfer function.

An example for the error calculation fitness function is the mean square error which can be calculated as follows:

$$f = \frac{1}{n}\sum_{i=1}^{n} f_i \tag{2.5}$$

The term f_i is calculated as the square of difference between the original step response and the reduced transfer function step response.

2.2 Genetic Algorithm Technique

The proposed model is able to reduce a transfer function of any order to a lower selected order. The inputs for the model are the original transfer function, number of iterations, desired reduction order, and generation population. The outputs are the reduced model transfer function, the mean square error, the time elapsed for reducing the transfer function, and the step and frequency responses of the models. Sometimes more time is needed for some transfer functions than the others to be reduced within the given criteria. The model contains a monitoring system for the step response and frequency response. The final implemented model is shown in Fig. 2.4.

The model is implemented using genetic operators: crossover and selection. The mutation operator is not used as it proves no extra improvement for the model. However, it adds more processing time. After generating an initial population from a selected domain, crossover operator is applied. Arithmetic crossover is used as an operator by generating two random numbers which add extra randomness and extend the domain of search of the genetic algorithm.

There is no physical need for a fixed domain of search for the coefficients of the transfer function. Selection operator is based on elitism where the fittest individuals in

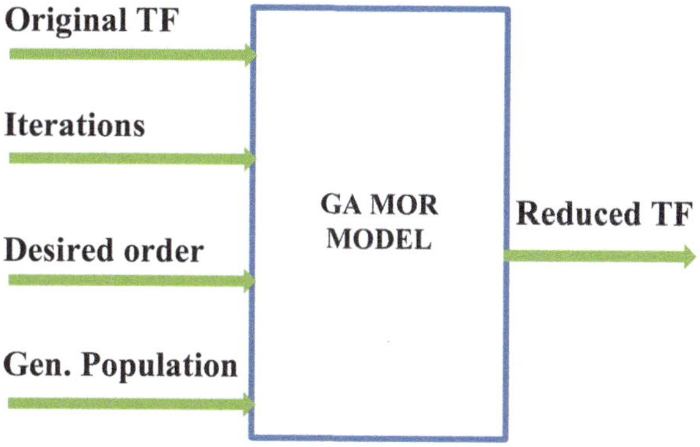

Fig. 2.4 The GA MOR model

the current population (with the least mean square error) are chosen and passed to the next generation. This method guarantees that the best individuals are passed to the next generation and at the same time it provides a high degree of randomness unlike the other methods. Other methods such as roulette wheel selection and methods based on giving weight to the individuals have a great disadvantage where individuals with higher weight may only pass to the next population which decrease randomness and decrease the probability of generating individuals of different properties. The chosen fitness function is the mean square error. The proposed methodology is shown in Fig. 2.5. A pseudo code for the implemented model that describes the procedures of implementation is given as follows in Listing 1:

Listing 1 Genetic algorithm's Pseudo-Code

Pseudo code of GA

```
    Input: problem, Population size, desired order
    Output: best_solution
    Create initial population
    Current population = initial population
    For (i = 1 to iterations)

        For (j = 1: to population size)
        Choose two random parents (current population)
        Do crossover producing two individuals
        End for

    Add best solution to best solutions set
    Choose the best individuals of new generation
    Current generation = best individuals
    End for
    If (iterations = max)
    Solution = best (best solutions)
    End if
```

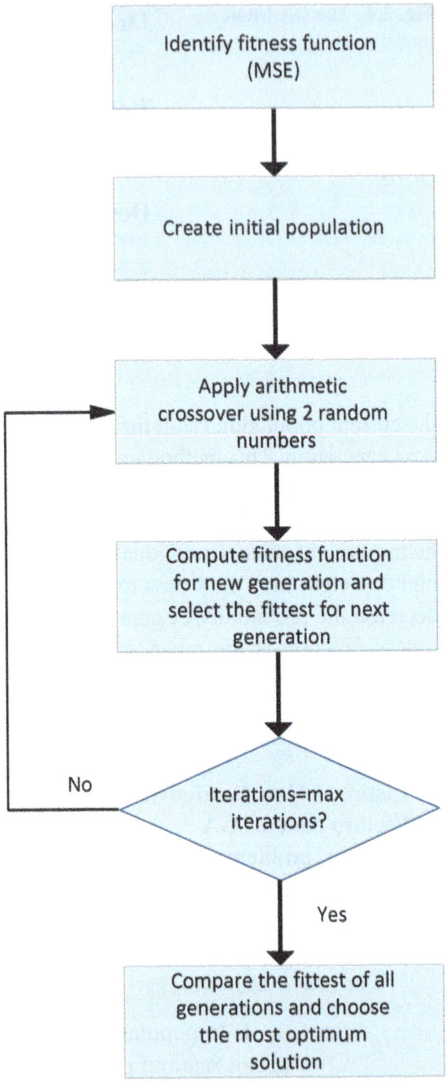

Fig. 2.5 GA-based MOR methodology

3 Testing of the Proposed Techniques

3.1 Setup of the Environment

The simulations are carried on Matlab 2012b software, core i7 processor, 6 GB Ram memory. The reduced order model is compared to the original one in terms of mean square error.

The testing procedures include a plot of step response for the reduced and the original TF, frequency response using Bode plot, mean square error, and the time elapsed.

The time taken for most of the original transfer functions to be reduced to lower order is nearly the same in all transfer functions, but sometimes some functions need more time than the others. The testing procedures are carried on more than 30 different transfer functions. The transfer functions have different forms of step responses and frequency responses where the step response of some of them has an exponential form and other responses were sinusoidal.

3.2 Genetic Algorithm Testing

The first transfer function to be tested is a tenth-order transfer function expressed as follows:

$$G_1(s) = \frac{\begin{array}{c} s^9 + 46.8s^8 + 957.6s^7 + 11144s^6 + 80511.9s^5 \\ +369601.6s^4 + 1060774.5s^3 + 1809006.4s^2 + 1669955.4s + 638266 \end{array}}{\begin{array}{c} s^{10} + 36.9s^9 + 620.8s^8 + 6257.9s^7 + 41888\,s^6 + 195879.7s^5 \\ +658023.2s^4 + 1611073.5s^3 + 2857356s^2 + 3425885.4s + 2110138.4 \end{array}} \tag{2.6}$$

The reduced second-order model is expressed as follows:

$$R_1(s) = \frac{2.574s + 1.847}{s^2 + 0.707s + 6.238} \tag{2.7}$$

The results of reducing the original model to a second-order model is shown in Table 2.3 and the step and frequency response are shown in Figs. 2.6 and 2.7, respectively.

The second transfer function is a tenth-order transfer function expressed as follows:

$$G_2(s) = \frac{\begin{array}{c} s^9 + 59s^8 + 1657.9s^7 + 28640.7\,s^6 + 334474.5s^5 + 2742122.4s^4 \\ +15867513.2s^3 + 62854022s^2 + 155453416.4s + 182004673.4 \end{array}}{\begin{array}{c} s^{10} + 49.5s^9 + 1168.6s^8 + 17109s^7 + 171437.4s^6 + 1227121s^5 + 6376179.3s^4 \\ +24002493s^3 + 64135934s^2 + 114485916.5s + 110910010.4 \end{array}} \tag{2.8}$$

The reduced second-order model is expressed as follows:

Table 2.3 The results of the reduced order model $G_1(s)$

Metric	Population	Iterations	MSE (%)	Time (sec)
$R_1(s)$	10	100	0.07	23

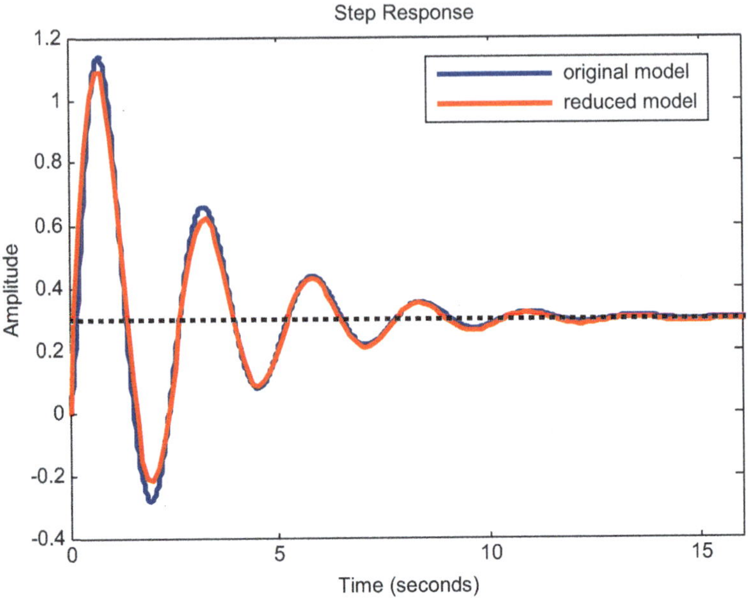

Fig. 2.6 Step response for the original and the reduced model of $G_1(s)$ with MSE 0.07%

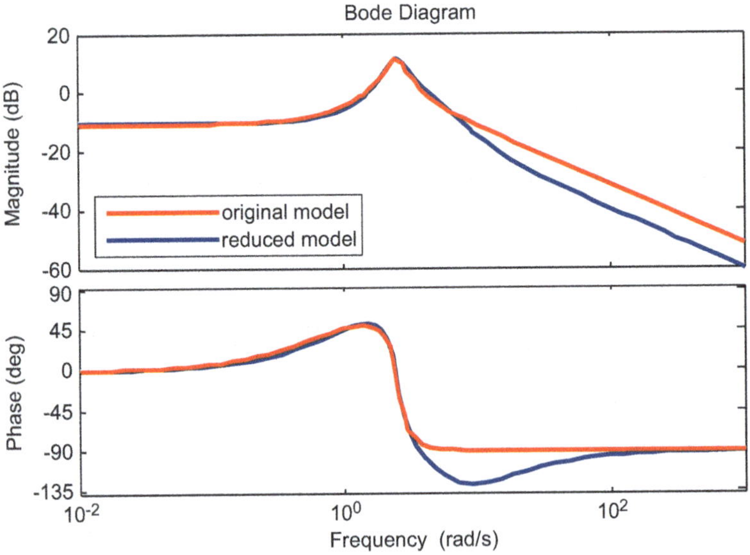

Fig. 2.7 The frequency response of the original and reduced transfer models for G1(s) with mag. MSE 0.77%

$$R_2(s) = \frac{1.167s + 16.69}{s^2 + 2.035s + 10.21} \tag{2.9}$$

The results of reducing the original model of the second transfer function to a second-order model is shown in Table 2.4 and the step and frequency response are shown in Figs. 2.8 and 2.9, respectively.

A comparison has been made between the results of the proposed model and the results of mathematical methods of Padé approximation, Routh approximation, and truncated-balanced method. The proposed model showed efficiency in terms of accuracy over Routh approximation while the results are nearly equal compared with Padé approximation, but the accuracy of Padé approximation is bad and sometimes unacceptable when testing transfer function of sinusoidal step response. Truncated-balanced method shows efficiency concerning accuracy over the proposed model, but the proposed model shows efficiency over it concerning frequency response as the frequency response of the truncated-balanced method is not accurate enough in most of the test cases and sometimes completely different from the original model. The first comparison is made between the results of the reduced order model of $G_1(s)$ and the mathematical methods. The step responses are shown

Table 2.4 The results of the reduced order model $G_2(S)$

Metric	Population	Iterations	MSE (%)	Time (sec)
$R_2(s)$	10	100	0.18	23

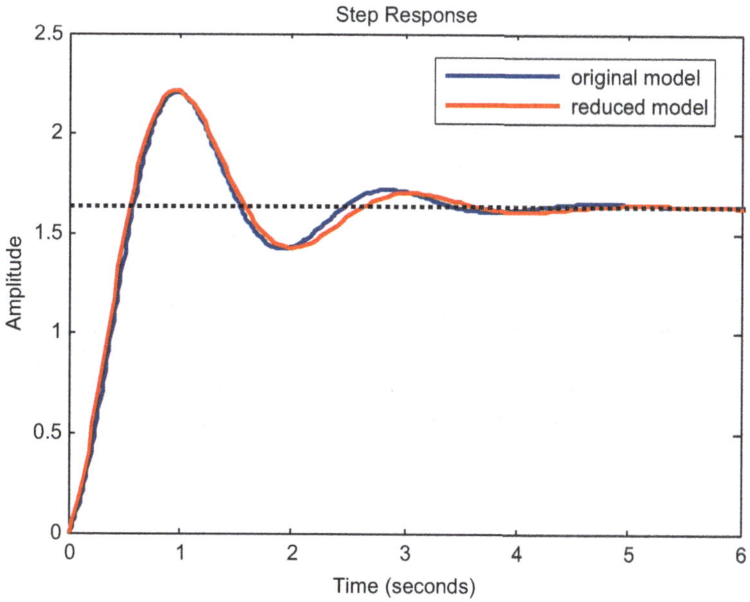

Fig. 2.8 Step response for $G_2(s)$ with MSE 0.18%

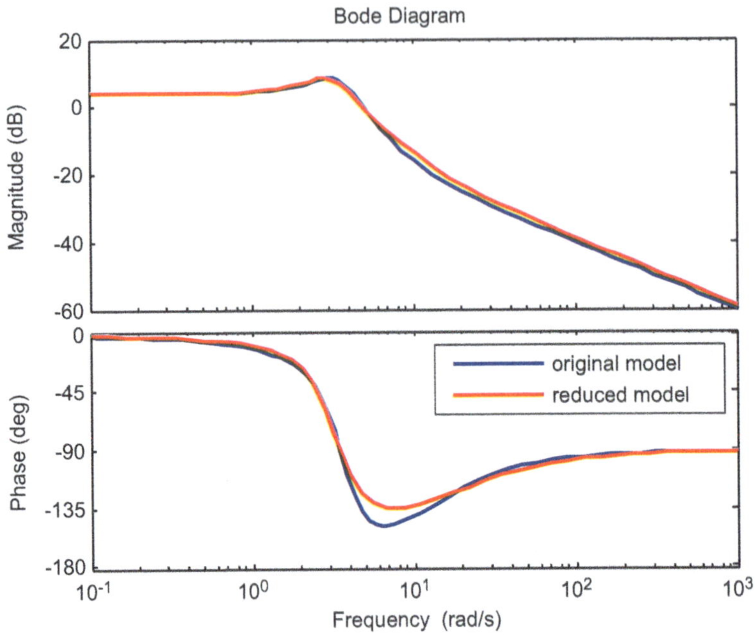

Fig. 2.9 The frequency response of the original and reduced transfer models for $G_2(s)$ with mag. MSE 0.50%

in Fig. 2.10 while the frequency responses are shown in Fig. 2.11. The results of the proposed model and the mathematical methods are compared in Table 2.5.

Although Padé approximation shows efficiency in terms of accuracy nearly similar to the efficiency of the proposed model, the step response is completely different from the step response of the original model in the first test case.

The second comparison is carried on $G_2(s)$ and the step responses are shown in Fig. 2.12 while the frequency responses are shown in Fig. 2.13. The results of the proposed model and the mathematical methods are compared in Table 2.6.

Compared to other related work that uses GA in MOR, the proposed model introduces a generic model achieving success in reducing any model of tenth order or less to any desired lower order. The simulations are carried on more than 30 models of different transfer function having exponential and sinusoidal forms of step responses including transfer functions having variety of complex poles while the previous trials are applied on specific models of fourth order in and eight order using simple schemes. Another advantage for the proposed model is that it can be used without defining a specific range of search for the solution and without additional procedures of gain adjusting and scaling.

The experiments compared the proposed model with powerful mathematical methods showing the efficiency of the proposed model over the mathematical model in terms of accuracy and preserving the properties of the original model. GA-based MOR is a promising solution in reducing the circuit-level simulation time.

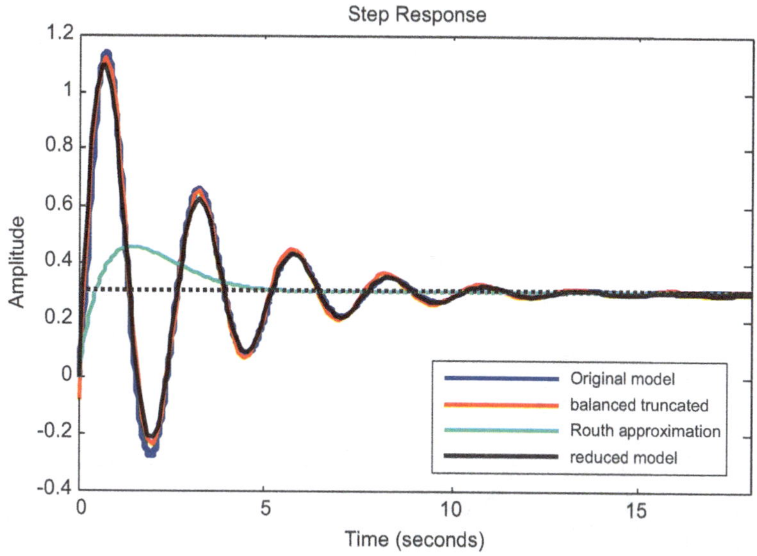

Fig. 2.10 The step response of the proposed model compared with the mathematical methods for $G_1(s)$

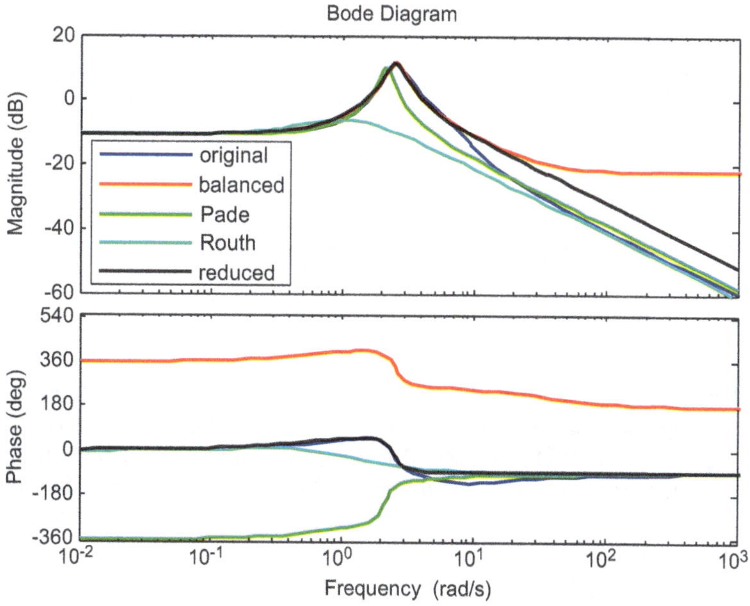

Fig. 2.11 Frequency responses for the first transfer function. Compared with the mathematical methods for $G_1(s)$

Table 2.5 A comparison of the results of all methods for the first transfer function

Metric	GA	Padé	Routh	Balanced
Step response MSE (%)	0.07	1760	5.28	0.05
Freq. response mag. MSE (%)	0.77	34	100	0.63
Freq. response phase MSE (%)	200	50	140	1.2×10^5
Time (s)	23	0.1	0.1	0.1

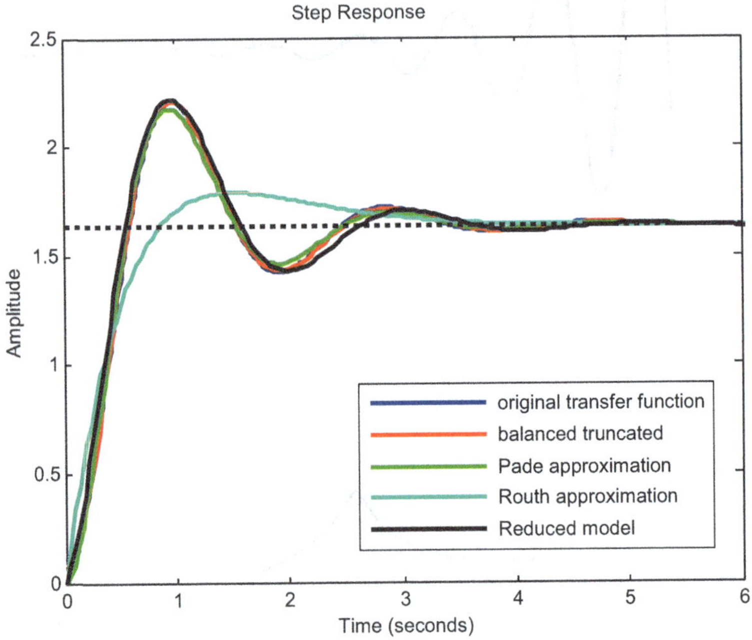

Fig. 2.12 Step responses for the second transfer function compared with the mathematical methods

4 Conclusions

This chapter presents a novel approach for MOR based on genetic algorithm. The proposed approach can obtain reduced order model out of a relatively complex model represented as a TF. This ROM is obtained in a reasonable time. Moreover, it has accepted error in time domain presented by the step response, and the least error in frequency domain presented by Bode plot compared to other famous conventional methods in MOR like moment matching based Padé approximation, Routh algorithm, or balanced truncation.

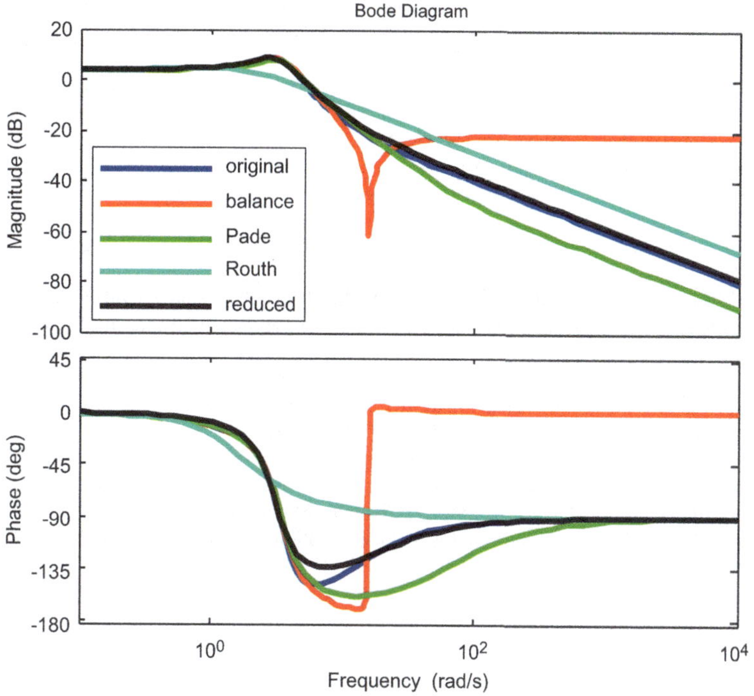

Fig. 2.13 frequency responses for the second transfer function compared with the mathematical methods

Table 2.6 A Comparison of the results for all methods for the second transfer function

Metric	GA	Padé	Routh	Balanced
Step response MSE (%)	0.07	1760	5.28	0.05
Freq. response mag. MSE (%)	0.50	0.33	23.8	0.19
Freq. response phase MSE	34	291	1×10^3	3.7×10^3
Time (s)	23	0.1	0.1	0.1

References

1. Y. Wu, L. Yu, W. Zhuang and J. Wang, *A Coverage-Driven Constraint Random-Based Functional Verification Method of Pipeline Unit*. Computer and Information Science, ACIS International Conference. (2009), pp. 1049–1054
2. M. Benjamin, D. Geist, A. Hartman, Y. Wolfsthal, G. Mas, R. Smeets, *A Study in Coverage-Driven Test Generation*. Design Automation Conference, 1999. Proceedings. 36th Issue. (1999), pp. 970–975
3. S. Fine and A. Ziv, *Coverage Directed Test Generation for Functional Verification Using Bayesian Networks*. Design Automation Conference, 2003. Proceedings Issue Date: 2–6 June, pp. 286–291

4. Z.S. Abo-Hammour, O.M.K. Alsmadi, and A.M. Al-Smadi, *Frequency-Based Model Order Reduction Via Genetic Algorithm Approach.* 7th International Workshop on Systems, Signal Processing and their Applications (WOSSPA), (2011)
5. Z.S. Abo-Hammour, O.M. Alsmadi, and A.M. Al-Smadi, *Frequency-Based Model Order Reduction Via Genetic Algorithm Approach.* 7th International Workshop on Systems, Signal Processing and their Applications (WOSSPA), 2011
6. I.Yun, L.A. Carastro, R. Poddar, M.A. Brooke, G.S. May, K.S. Hyun, and K.E. Pyun, Extraction of passive device model parameters using genetic algorithms" ETRI J.. **22**(1), (2000)
7. K. Thirugnanam, E. Reena, M. Singh, P. Kumar, Mathematical modeling of Li-Ion battery using genetic algorithm approach for V2G applications. IEEE Trans. Energy Conv. **29**(2), (2014)
8. Z.S. Abo-Hammour, O.M. Alsmadi, and A.M. Al-Smadi, *Frequency-Based Model Order Reduction via Genetic Algorithm Approach*, 7th International Workshop on Systems, Signal Processing and their Applications (WOSSPA), (2011)
9. G. Parmar, M.K. Pandey, V. Kumar, *System Order Reduction using GA for Unit Impulse Input and A Comparative Study using ISE & IRE*, International Conference on Advances in Computing, Communication and Control (ICAC3'09), (2009)
10. I. Yun, L.A. Carastro, R. Poddar, M.A. Brooke, G.S. May, K.S. Hyun, and K.E. Pyun, Extraction of passive device model parameters using genetic algorithms. ETRI J.. **22**(1), (2000)
11. K. Thirugnanam, E. Reena, M. Singh, P. Kumar, Mathematical modeling of Li-Ion battery using genetic algorithm approach for V2G applications. IEEE Trans. Energy Conv. **29**(2), (2014)

Chapter 3
Thermo-Inspired Machine Learning Algorithm: Simulated Annealing

1 Introduction

Simulated Annealing (SA) is a global optimization algorithm. It belongs to the stochastic optimization algorithms. The idea of SA was published in a paper by Metropolis [1], this idea was inspired by the annealing process in metallurgy. In this process, a material is heated to a very high temperature so that the particles of the material are freely moving, and then the material is gradually cooled under certain conditions so that the particles begin to take the narrower paths till they collectively reach the minimum energy state level in the material as depicted in Fig. 3.1. In other words, it is a physical phenomenon that occurs when the metal is heated at a very high temperature, and then slowly cooled [2–9]. By analogy with this physical process, each step of the SA algorithm attempts to replace the current solution by a random solution till the desired output is obtained. At each step of the algorithm, SA decides between moving the system to a new state and staying in the current state. This decision is done probabilistically. The analogy between simulated annealing and physical system can be shown in Table 3.1.

2 The Proposed Technique

2.1 Problem Formulation

Consider we have a linear system of nth order of q inputs and r outputs described using state space in time domain as follows:

$$H = Ax(t) + Bu(t) \tag{3.1}$$

$$Y(t) = Cx(t) \tag{3.2}$$

K. S. Mohamed, *Machine Learning for Model Order Reduction*,
https://doi.org/10.1007/978-3-319-75714-8_3

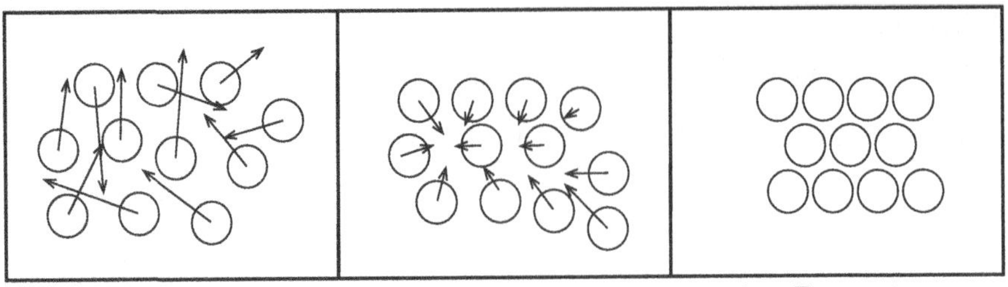

Fig. 3.1 Molecules movement as a function of temperature

Table 3.1 The analogy between simulated annealing and physical system

Physical system	SA
State	Solution
Energy	Cost function
Temperature	Control parameter
Ground state	Optimum solution

where H is n-dimensional state vector, u is q-dimensional control vector, and Y is r-dimensional output vector with $q \leq n$ and $r \leq n$. Also, A is $n \times n$ system matrix, B is $n \times q$ input matrix, and C is $r \times n$ output matrix.

By converting the state space model from time domain to frequency domain, the transfer function can be described in Eq. (3.4):

$$G(s) = \frac{N(s)}{D(s)} = \frac{\sum_{i=0}^{n-1} A_i s^i}{\sum_{i=0}^{n} a_i s^i} \tag{3.3}$$

where $N(s)$ is the numerator and $D(s)$ is the denominator of the higher order system. Also, A_i and a_i are the coefficients of numerator and denominator, respectively.

The problem is to find a mth lower order model $R^m(s)$, where $m < n$ in the following form:

$$R(s) = \frac{N^m(s)}{D^m(s)} = \frac{\sum_{i=0}^{n-1} B_i s^i}{\sum_{i=0}^{n} b_i s^i} \tag{3.4}$$

In addition, we must take into consideration that the reduced model maintains the characteristics of the original transfer function.

2.2 *Simulated Annealing Technique*

SA is a probabilistic algorithm used to find the optimum solution of problems that have multiple local minima. The algorithm's plan is to probabilistically rearrange the problem's solution space in a way that enables it to accept samples that could have lower cost function, but this is for the sake of finding the global minima.

This decision is based on a probabilistic function that is inspired from thermodynamics; this function will be introduced in the pseudo code of the algorithm, and the procedures of SA algorithm. The procedures of SA are much more similar to what occurs in the annealing process in metals. By analogy, it is clear that free particles are equivalent to free variables, and energy of the particles is equivalent to the cost function that we want to optimize.

The following procedures describe the behavior of SA algorithm, and how it works till finding the optimum solution for an optimization problem.

(a) Initialize a high temperature.
(b) Create an initial solution.
(c) Asses that solution's cost.
(d) Modify to get neighboring solution.
(e) Asses that new solution's cost.
(f) Stochastically select a solution by using the probabilistic **function e** $^{(-\Delta E/T)}$. Where ΔE is the difference between the current solution and the neighborhood new solution (decreasing factor), and T is the current temperature value.
(g) Reduce temperature with decreasing factor.
(h) Repeat procedures from (d to g) till finding the optimum solution.

By studying the stochastic function $e^{(\Delta E/T)}$, we notice that by decreasing the temperature, the probability of choosing new values that might get a better solution compared to the last local minimum founded in the solution space will be decreased. The local minimum becomes the global one in the solution space at a very low temperature. This means that finding the global minimum—optimum solution—is getting closer by decreasing the temperature.

The pseudo-code in Listing 1 describes in detail the procedures of SA algorithm in finding the optimum solution that we are seeking for.

SA algorithm is designed for solving discrete optimization problems, though it has been used for continuous function optimization problems.

The convergence proof presented in suggests that with a slow enough cooling schedule, the algorithm will always converge to the global minimum. The drawback of this is that the number of samples chosen from the solution space might exceed the total number of samples, and the run-time of the algorithm will be increased.

We introduce a novel model that reduces the order of relatively complex TFs to lower order ones using SA algorithm. Our model is generic and user-friendly. The generality feature is considered the most powerful feature in our model. As shown in Fig. 3.2, it has four inputs and four outputs, the user should enter the following parameters as inputs:

1. Original high order TF presented as numerator and denominator coefficients' in vector form.
2. Desired order of the reduced order model.
3. Number of iterations of the algorithm.

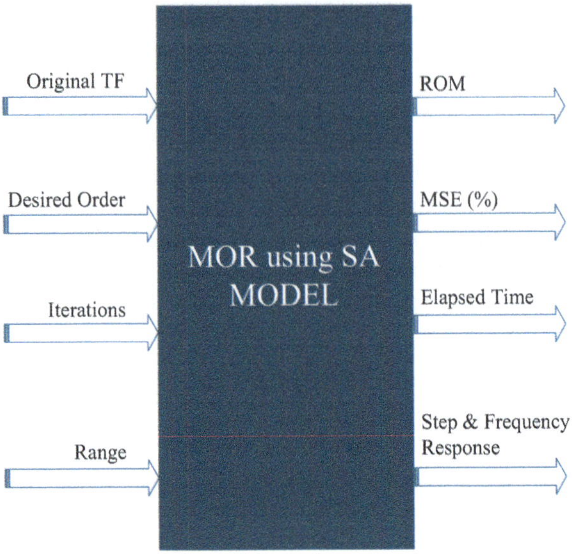

Fig. 3.2 MOR using SA proposed model

4. Range for the solution space where the algorithm should search. This range consists of two ranges, one for the numerator coefficients, and the other one is for denominator coefficients.

The outputs of our model are:

1. Reduced order Model.
2. Error between original TF and reduced TF.
3. Time Elapsed for the MOR reduction process.
4. Graph showing comparison between original TF, and reduced order TF in: Frequency response and unit step time response.

Listing 1 Simulated Annealing algorithm's Pseudo-Code
Pseudo-code: Simulated annealing algorithm

```
Input: ProblemSize, iterations, MaxTemperature, CurrentTemperature
Output: BestSolution
CurrentSolution ← CreateInitialSolution(ProblemSize)
BestSolution ← CurrentSolution
For(i = 1 to iterations)
NewSolution ← createNeighborSolution(CurrentSolution)
CurrentTemperature ← MaxTemperature*DecreasingFactor
If (cost(NewSolution) ≤ cost(CurrentSolution))
CurrentSolution ← NewSolution
If (cost(CurrentSolution) ≤ cost(BestSolution))
```

(continued)

```
BestSolution ← CurrentSolution
End if
ElseIf (e^(CurrentSolution-NewSolution/T) > Rand())
CurrentSolution ← NewSolution
End if
End for
Return (BestSolution)
```

3 Testing of the Proposed Techniques

3.1 *Setup of the Environment*

The simulations are carried on Matlab 2012b software, core i7 processor, 6 GB Ram memory. The reduced order model is compared to the original one in terms of mean square error.

The testing procedures includes a plot of step response for the reduced and the original TF, frequency response using Bode plot, mean square error, and the time elapsed.

The time taken for most of the original transfer functions to be reduced to lower order is nearly the same in all transfer functions, but sometimes some functions need more time than the others. The testing procedures were carried on more than 30 different transfer functions. The transfer functions have different forms of step responses and frequency responses where the step response of some of them has an exponential form and other responses were sinusoidal.

3.2 *Simulated Annealing Testing*

In this subsection, we present out the results for our SA model tested on some samples of transfer functions and we compare between them and conventional methods.

Example 1 The following TF represents a tenth-order system; this system is simplified by our model to be a second-order system; this ROM will be presented in Table 3.3.

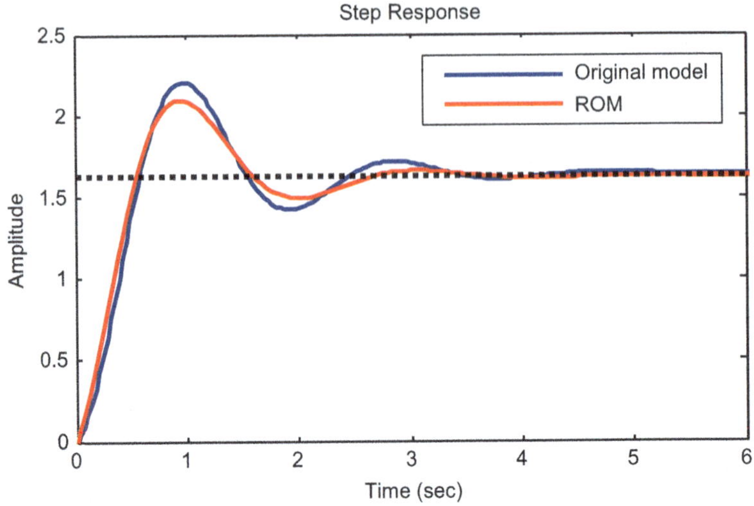

Fig. 3.3 Unit step time response of TF (4) versus its ROM with MSE = 0.32%

$$H(s)=\frac{\begin{array}{c}(S^9+59S^8+1657.9S^7+28640.7S^6+334474.5S^5+2742122.4S^4\\+15867513.2S^3+62854022S^2+155453416.4S+182004673.4)\end{array}}{\begin{array}{c}(S^{10}+49.5S^9+1169S^8+17109S^7+171437.4S^6+1227121S^5+6376179S^4\\+24002493S^3+64135934S^2+114485916.5S+110910010)\end{array}} \tag{3.5}$$

Figures 3.3 and 3.4 illustrate comparisons between the original system and the ROM in terms of unit step time domain response, and frequency response.

Example 2 The following TF represents a tenth-order system; this system is simplified by our model to be a second-order system.

$$H(s)=\frac{\begin{array}{c}(S^9+46.8S^8+957.6S^7+11144S^6+80511.9S^5+369601.6S^4\\+1060774.5S^3+1809006.4S^2+1669955.4S+638266)\end{array}}{\begin{array}{c}(S^{10}+36.9S^9+620.8S^8+6257.9S^7+41888S^6+195879.7S^5+658023.2S^4\\+1611073.5S^3+2857356S^2+3425885.4S+2110138.4)\end{array}} \tag{3.6}$$

Figures 3.5 and 3.6 illustrate comparisons between the original system and the ROM in terms of unit step time domain response, and frequency response.

Comparing the results with the output of some conventional MOR methods such as moment-matching-based Padé approximation method, Routh Approximation, and balanced truncation method are shown in Figs. 3.7 and 3.8 and Table 3.2 for Example 1 and shown in Figs. 3.9, 3.10, and 3.11 for Example 2. The final performance results are summarized in Table 3.3.

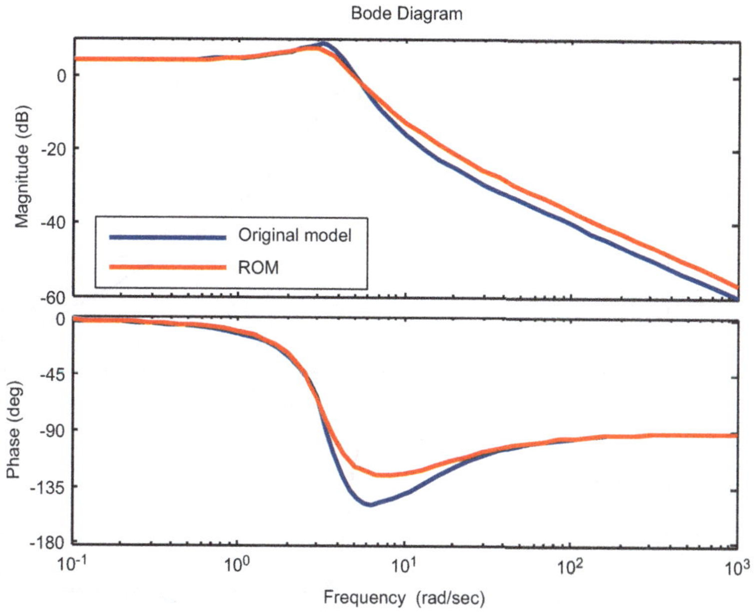

Fig. 3.4 Frequency response of TF (4) versus its ROM with MSE = 1.6% in magnitude, and 0.01% in phase

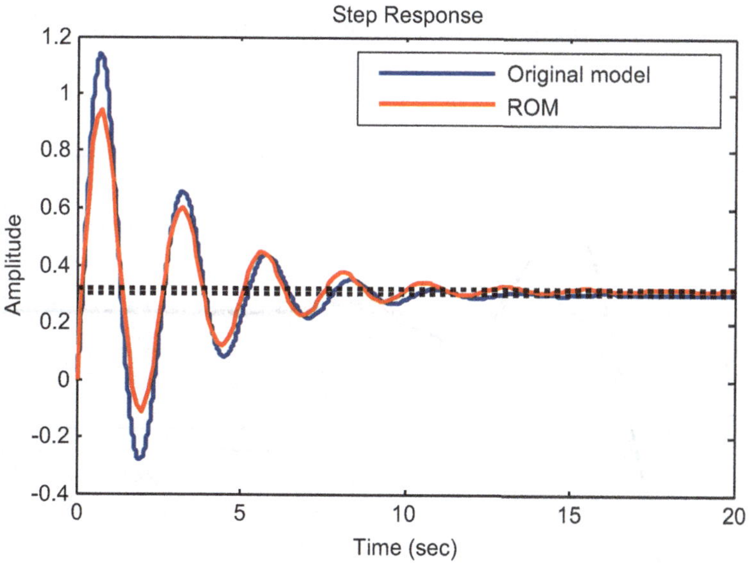

Fig. 3.5 Unit step time response of TF (6) versus its ROM, with MSE = 0.35%

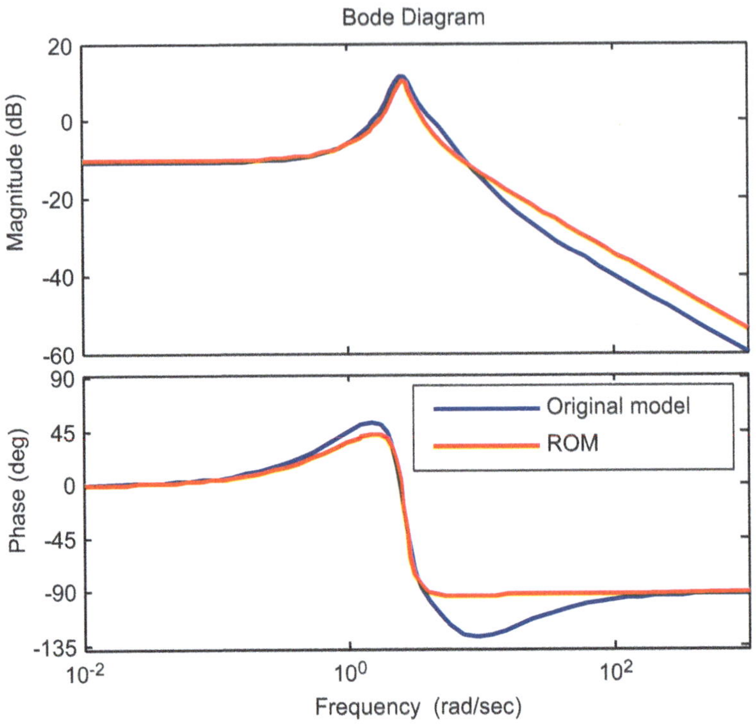

Fig. 3.6 Frequency response of TF (6) versus its ROM, with MSE = 6% in magnitude, and 0.02% in phase

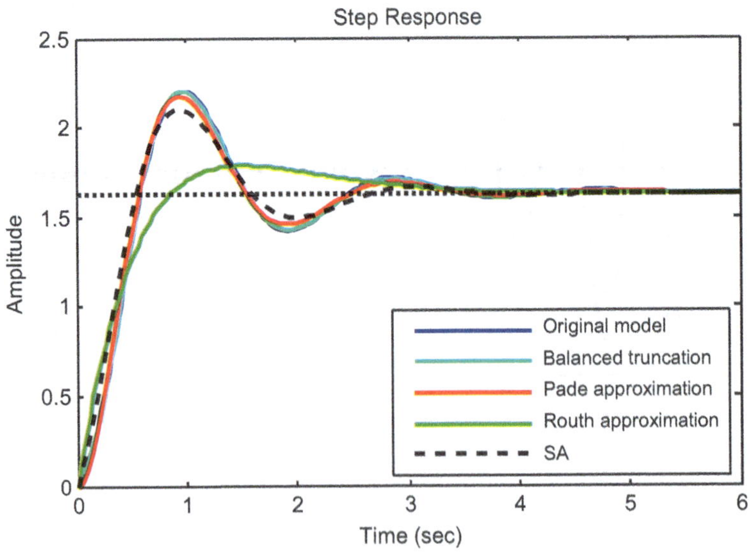

Fig. 3.7 Unit step time response comparison among MOR conventional methods and the proposed method

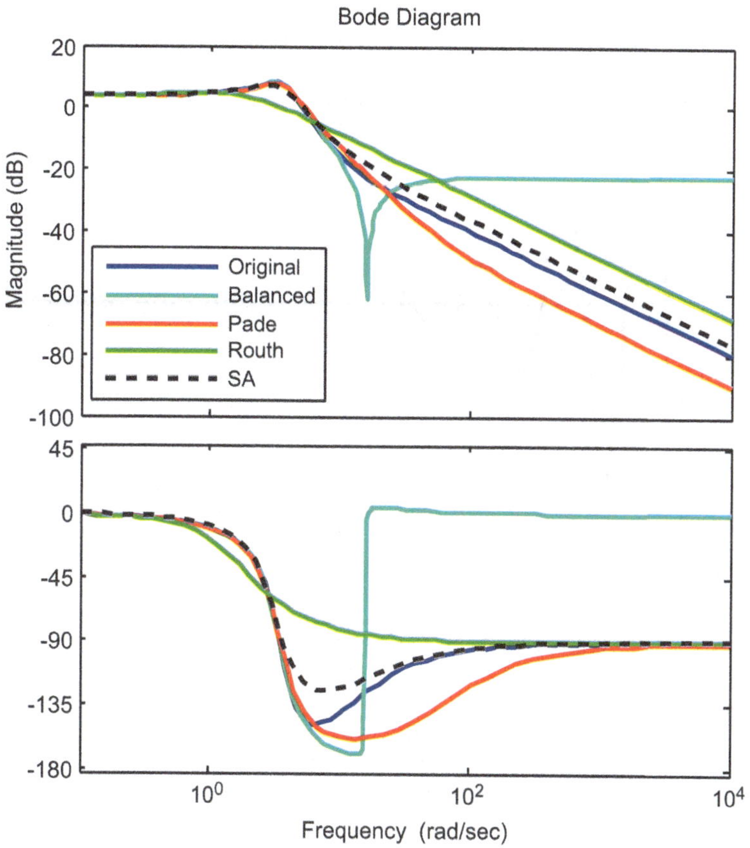

Fig. 3.8 Frequency response comparison among MOR conventional methods and the proposed method in Example 1

Table 3.2 A Comparison among conventional MOR methods and SA applied on Example 1

Metrics	SA	Padé	Routh	Balanced
Frequency domain MSE (%) (magnitude)	1.6	0.3	24	0.2
Frequency domain MSE (%) (phase)	0.01	0.03	0.01	0.4
Time domain MSE (%)	0.32	0.05	4.1	0.01
Elapsed time (s)	56	0.1	0.1	0.1

4 Conclusions

This chapter presents a novel approach for MOR based on simulated annealing. Our approach aims at obtaining a reduced order model out of a relatively complex model represented as a TF. This ROM is obtained in a reasonable time. Moreover, it has accepted error in time domain presented by the step response, and the least error in

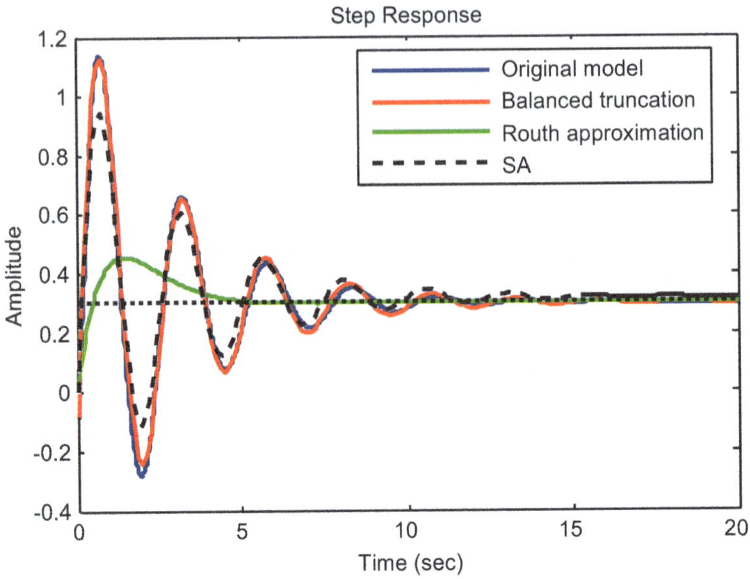

Fig. 3.9 Unit step time response comparison among MOR conventional methods and the proposed method except for Padé approximation in Example 2

Fig. 3.10 Unit step time response comparison between Example 2 and Padé approximation. It shows a clear drawback of Padé approximation method

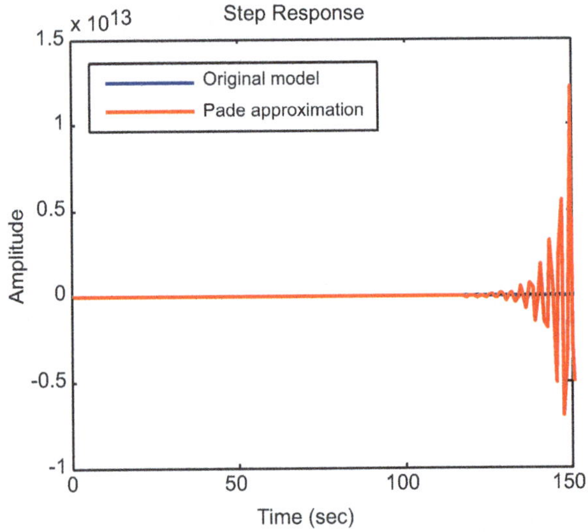

frequency domain presented by Bode plot compared to other famous conventional methods in MOR like moment matching based Padé approximation, Routh algorithm, or balanced truncation.

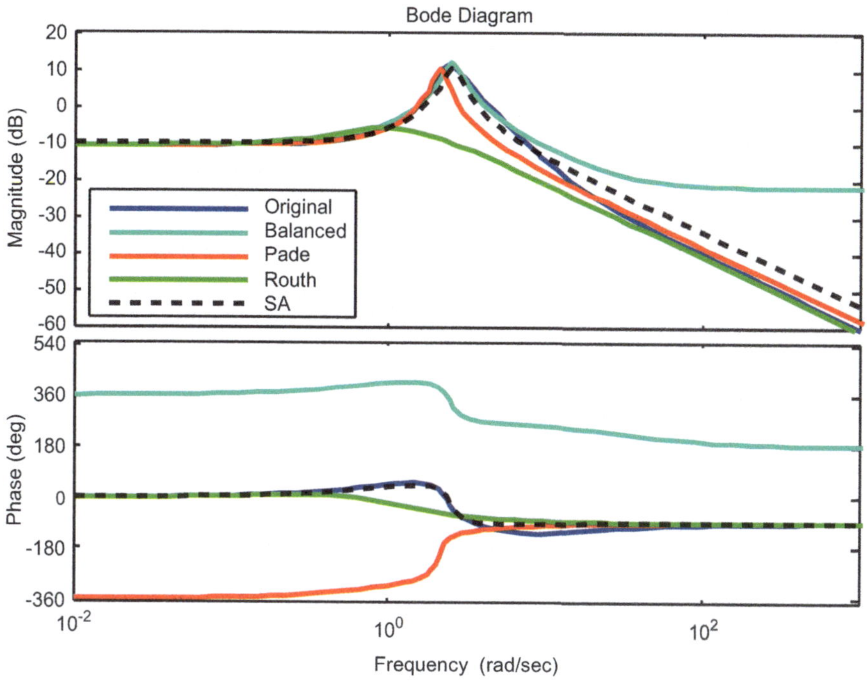

Fig. 3.11 Frequency response comparison among MOR conventional methods and the proposed method including Padé approximation method in Example 2

Table 3.3 A comparison among conventional MOR methods and SA applied on Example 2 show that our method has the best frequency response accuracy

Metrics	SA	Padé	Routh	Balanced
Frequency domain MSE (%) (magnitude)	6	34	100	0.6
Frequency domain MSE (%) (phase)	0.02	5.7	0.01	12
Time domain MSE (%)	0.35	1.7	5.35	0.05
Elapsed time (s)	33	0.2	0.1	0.5

References

1. N. Metropolis, W. Arianna, M. Rosenbluth, H. Augusta, Equation of state calculations by fast computing machines. J. Chem. Phys. **21**(6), 1087 (1953)
2. S. Kirkpatrick, C.D. Gelatt Jr., M.P. Vecchi, Optimization by simulated annealing. Science **220** (4598), 671–680 (1983)
3. V. Černý, Thermodynamical approach to the traveling salesman problem: an efficient simulation algorithm. J. Optim. Theory Appl. **45**, 41–51 (1985)
4. S. Geman, D. Geman, Stochastic relaxation, Gibbs' distribution and the Bayesian restoration of images. IEEE Trans. Pattern Anal. Mach. Intell. **6**, 721–741 (1984)

5. B. Gidas, Nonstationary Markov chains and convergence of the annealing algorithm. J. Statist. Phys. **39**, 73–131 (1985)
6. R.A. Holley, S. Kusuoka, D.W. Stroock, Astmptotics of the spectral gap with applications to theory of simulated annealing. J. Funct. Anal. **83**, 333–347 (1989)
7. G.C. Jeng, J.W. Woods, Simulated annealing in compound Gaussian random fields. IEEE Trans. Inform. Theory **36**, 94–107 (1990)
8. Kuchner, H. J.. Asymptotic global behavior for stochastic approximations and diffusions with slowly decreasing noise effects: global minimization via Monte Carlo. Technical Report LCDS 85-7, Dept. Applied Mathematics, Brown Univ, (1985)
9. V. Granville and M. Krivanek and J–P. Rasson, Simulated annealing: a proof of convergence, *IEEE Trans. Pattern Anal. Mach. Intell.*, (1994)

Chapter 4
Nature-Inspired Machine Learning Algorithm: Particle Swarm Optimization, Artificial Bee Colony

1 Introduction

The PSO method is a member of wide category of swarm intelligence methods for solving the optimization problems. Particle swarm optimization technique is computationally effective and easier.

Inspired by the flocking of birds or the schooling patterns of fish, Particle Swarm Optimization (PSO) was invented by Russell Eberhart and James Kennedy in 1995 [1]. Originally, these two started out developing computer software simulations of birds flocking around food sources, and then later realized how well their algorithms worked on optimization problems [2–6] .

Suppose the following scenario: a group of birds is randomly searching food in an area. There is only one piece of food in the area being searched. All the birds do not know where the food is. But they know how far the food is in each iteration. So what's the best strategy to find the food? ... The effective one is to follow the bird which is nearest to the food. So, the one who is closest to the food chirps the loudest and the other birds swing around in his direction. If any of the other circling birds comes closer to the target than the first, it chirps louder and the others veer over toward him. This tightening pattern continues until one of the birds happens upon the food. It's an algorithm that's simple and easy to implement.

In PSO, each single solution is a "bird" in the search space. We call it "particle." All of particles have fitness values which are evaluated by the fitness function to be optimized, and have velocities which direct the flying of the particles. The particles fly through the problem space by following the current optimum particles.

The ant colony optimization algorithm (ACO) was proposed by Dorigo et al. [7]. The idea of ant colony algorithms comes from biology and from the observation of ant social behavior. Ant colony algorithm is the simulation of cooperation of real ant group. Each ant searches the solution independently in the candidate solution space, meanwhile leaves some information on the found solution. This is done using pheromone trails, which ants deposit whenever they travel, as a form of indirect

K. S. Mohamed, *Machine Learning for Model Order Reduction*,
https://doi.org/10.1007/978-3-319-75714-8_4

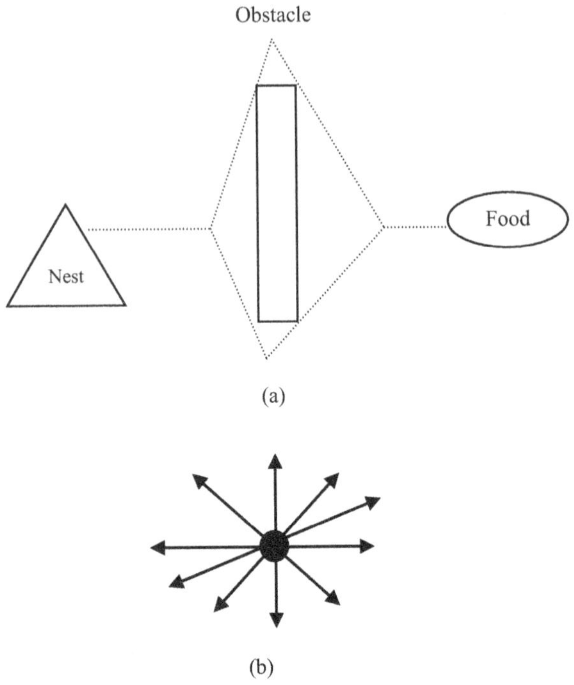

Fig. 4.1 (**a**) Ant colony: Shortest path to food source. If two ants set out at the same time, one taking route A and one route B, which is longer, then the ant taking A will have travelled back and forth between the food source twice in the same time that the other ant has travelled back and forth once. Therefore there will be a stronger pheromone trail on route A compared to route B. (**b**) global directions search

communication. Ants leave more information on the solution with better performance, which has more possibility to be selected. All the solutions have the same information on it in the elementary period of the algorithm. With the progress of the computing, the better solution gets more and more information. Therefore, the algorithm reaches the optimization or approximately optimization solution [8, 9].

As shown in Fig. 4.1a when ants leave their nest to search for a food source, they randomly rotate around an obstacle, and initially the pheromone deposits will be the same for the right and left directions. The direction from the beginning is random as depicted in Fig. 4.1b. When the ants in the shorter direction find a food source, they carry the food and start returning back, following their pheromone trails, and still depositing more pheromone. An ant will most likely choose the shortest path when returning back to the nest with food as this path will have the most deposited pheromone. For the same reason, new ants that later starts out from the nest to find food will also choose the shortest path. Over time, this positive feedback process prompts all ants to choose the shorter path (Fig. 4.1).

Artificial bee colony (ABC) algorithm is based on swarm intelligence where it utilizes the foraging behavior of honey bee's colonies. This model has three main components: employed bees, unemployed bees, and food sources [10–14]. Unemployed bees are divided into two groups, which are scout and onlooker. ABC algorithm is workings as follows:

- Scout bees randomly start the food-searching process without any guidance.
- Nectar is moved to the hive from discovered food source by scout bees.
- Employed bees bring food to the hive, go back to the source or nectar source information is transferred to scout bees in hive via information dance.
- Employed bee which consumes food, starts to work as a scout bee again and it searches the new food sources.
- Information of food sources depending on the quality and frequency of the dance from employed bees is obtained by scout bees in the hive, and they are canalized to the food sources.

The main advantage of this algorithm is that it has few control parameters. In this work, the colony size is 100 bees.

2 The Proposed Technique

2.1 *Statement of the Problem*

Consider we have a linear system of nth order of q inputs and r outputs described using state space in time domain as follows:

$$H = Ax(t) + Bu(t) \tag{4.1}$$

$$Y(t) = Cx(t) \tag{4.2}$$

where H is n-dimensional state vector, u is q-dimensional control vector, and Y is r-dimensional output vector with $q \leq n$ and $r \leq n$. Also, A is $n \times n$ system matrix, B is $n \times q$ input matrix, and C is $r \times n$ output matrix.

By converting the state space model from time domain to frequency domain, the transfer function can be described in Eq. (4.3):

$$G(s) = \frac{N(s)}{D(s)} = \frac{\sum_{i=0}^{n-1} A_i s^i}{\sum_{i=0}^{n} a_i s^i} \tag{4.3}$$

where $N(s)$ is the numerator and $D(s)$ is the denominator of the higher order system. Also, A_i and a_i are the coefficients of numerator and denominator, respectively.

The problem is to find a mth lower order model $R^m(s)$, where $m < n$ in the following form:

$$R(s) = \frac{N^m(s)}{D^m(s)} = \frac{\sum_{i=0}^{n-1} B_i s^i}{\sum_{i=0}^{n} b_i s^i} \tag{4.4}$$

In addition, we must take into consideration that the reduced model maintains the characteristics of the original transfer function.

2.2 Particle Swarm Optimization Technique

PSO is initialized with a group of random particles (solutions) and then searches for optima by updating generations. In every iteration, each particle is updated by following two "best" values. The first one is the best solution (fitness) each particle has achieved so far, this value is called Pbest. Another "best" value that is tracked by the particle swarm optimizer is the best value, obtained so far by any particle in the population. This best value is a global best and called Gbest. Each particle consists of: Data representing a possible solution, a velocity value indicating how much the Data can be changed, a personal best (Pbest) value indicating the closest the particle's Data has ever come to the Target.

The particles' data could be anything. In the flocking birds' example above, the data would be the *X, Y, Z* coordinates of each bird. The individual coordinates of each bird would try to move closer to the coordinates of the bird which is closer to the food's coordinates (Gbest). If the data is a pattern or sequence, then individual pieces of the data would be manipulated until the pattern matches the target pattern.

The velocity value is calculated according to how far an individual's data is from the target. The further it is, the larger the velocity value. In the birds' example, the individuals furthest from the food would make an effort to keep up with the others by flying faster toward the Gbest bird. If the data is a pattern or sequence, the velocity would describe how different the pattern is from the target, and thus how much it needs to be changed to match the target.

Each particle's Pbest value only indicates the closest the data has ever come to the target since the algorithm started.

After first initialization of the two best values, the particle updates its velocity and positions each iteration with the following equations (velocity and position update equations):

$$v_{i+1} = w * v_i + c_1 * \text{rand}\ (i) * (\text{Pbest}_i - \text{present}_i) + c_2 * \text{rand}\ (i) * (\text{Gbest}_i - \text{present}_i) \tag{4.5}$$

$$\text{present}_{i+1} = \text{present}_i + v_i \tag{4.6}$$

where

- v_i is the velocity of the current particle at this iteration.
- Present_i is the current particle (solution).
- w = inertia weight factor.
- c_1, c_2 = cognitive and social acceleration factors, respectively.
- rand () = random numbers uniformly distributed in the range (0, 1).
- Pbest = the best value obtained by the current particle till now.
- Gbest = Global best of the group.

Pseudo code of Particle Swarm Optimization Algorithm is shown below in Listing 4.1 and its flow chart is shown in Fig. 4.2. The Algorithmic steps involved in PSO Algorithm are as follows:

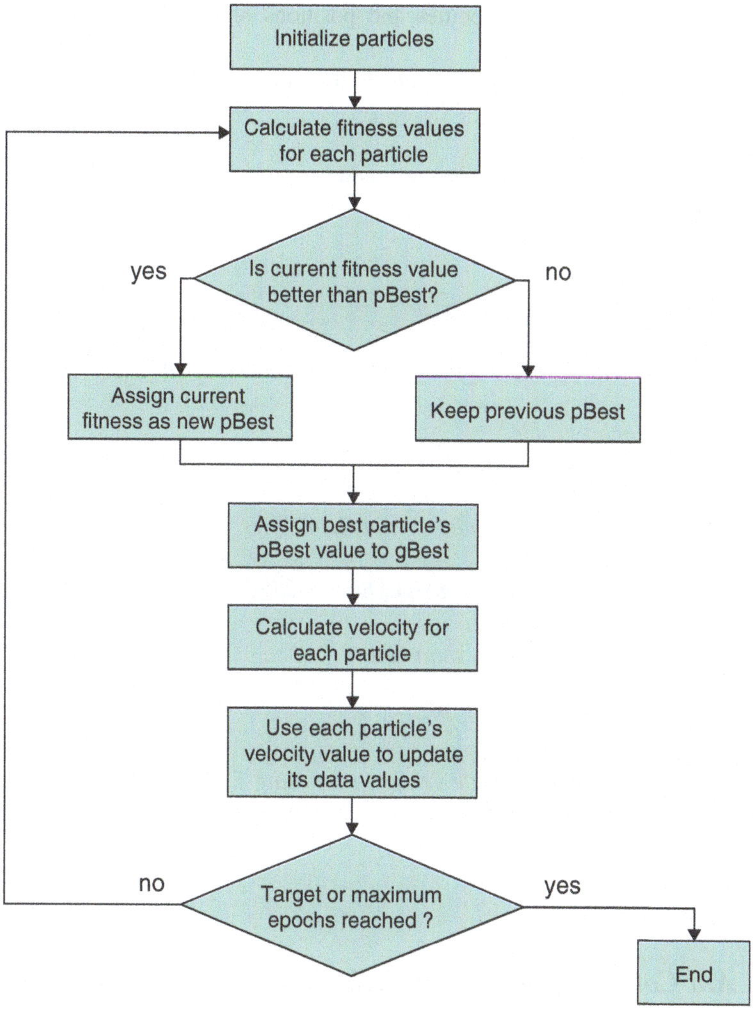

Fig. 4.2 PSO flow chart for model order reduction

Step 1: Select the parameters of PSO.
Step 2: Initialize a Population of particles with random positions and velocities in the problem space.
Step 3: Evaluate the desired optimization fitness Function for each particle.
Step 4: For each Individual particle, compare the Particles fitness value with its Pbest. If the Current value is better than the Pbest value, then update Pbest for agent *i*.
Step 5: Identify the particle that has the best fitness Value. The value of its fitness function is identified a Gbest.

Step 6: Compute the new velocities and positions of the particles according to Eqs. (4.5) and (4.6).
Step 7: Repeat steps 3–6 until the stopping criterion of maximum generations is met.

Listing 4.1 PSO algorithm's pseudo-code
Pseudo-code: PS algorithm

```
For each particle
{
    Initialize particle
}
Do until maximum iterations or minimum targeted error
{
    For each particle
    {
        Calculate Data fitness value
        If the fitness value is better than Pbest
        {
            Set Pbest = current fitness value
        }
        If Pbest is better than Gbest
        {
            Set Gbest = Pbest
        }
        Calculate particle velocity according Eq. (4.1)
        Update particle position according Eq. (4.2)
    }
}
```

2.3 Ant Colony Optimization Technique

The idea is to generate several random moves and see which one is the best among them. The proposed algorithm can be summarized in the following steps:

1. Initialize the nest structure by generating random starting search direction vectors.
2. Associate each direction with a pheromone variable initialized at the beginning of each iteration.
3. Define the search radius which determines the maximum distance that an ant can move at a single time.
4. Send ants in various search directions.

5. Calculate the cost function (Mean square error) when the ant reach the maximum distance and back again to the nest. Continue until reaches starting node as finished "tour" is considered a solution.
6. Re-tune the solution and update pheromone: pheromone_new = pheromone_old + Cost.
7. Repeat until you obtain the required MSE/the best solution.

2.4 Artificial Bee Colony Optimization Technique

ABC algorithm is workings as follows:

1. Scout bees randomly start the food-searching process without any guidance.
2. Nectar is moved to the hive from discovered food source by scout bees.
3. Employed bees bring food to the hive, go back to the source or nectar source information is transferred to scout bees in hive via information dance.
4. Employed bee which consumes food, starts to work as a scout bee again and it searches the new food sources.
5. Information of food sources depending on the quality and frequency of the dance from employed bees is obtained by scout bees in the hive, and they are canalized to the food sources.

3 Testing of the Proposed Techniques

3.1 Setup of the Environment

The simulations are carried on Matlab 2012b software, core i7 processor, 6 GB Ram memory. The reduced order model is compared to the original one in terms of mean square error.

The testing procedures include a plot of step response for the reduced and the original TF, frequency response using Bode plot, mean square error, and the time elapsed.

The time taken for most of the original transfer functions to be reduced to lower order is nearly the same in all transfer functions, but sometimes some functions need more time than the others. The testing procedures were carried on more than 30 different transfer functions. The transfer functions have different forms of step responses and frequency responses where the step response of some of them has an exponential form and other responses were sinusoidal.

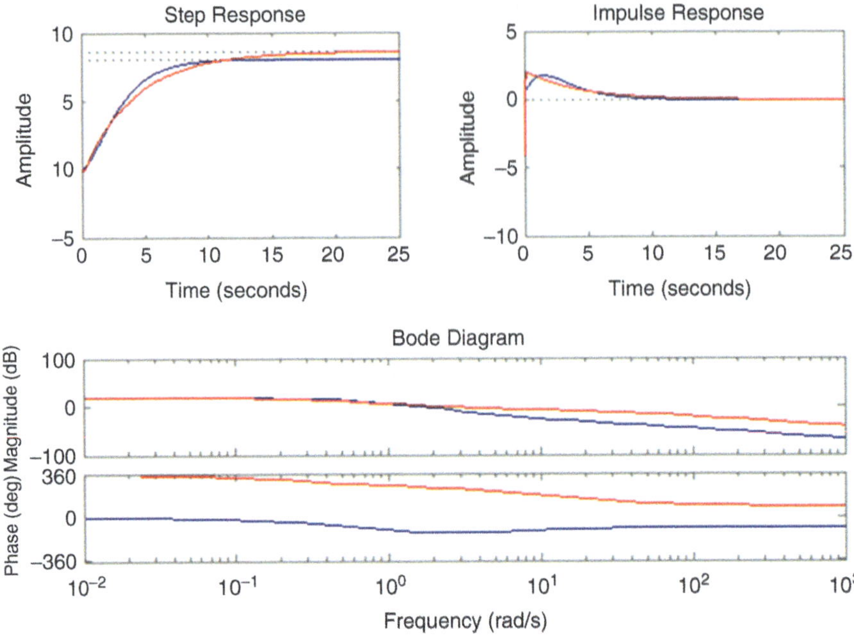

Fig. 4.3 Step response and frequency response comparison between original and reduced transfer function (Example 1)

3.2 Particle Swarm Optimization Testing

In this subsection, we show some examples of the experimental transfer functions, and the result of reduction. The examples are shown in Figs. 4.3 and 4.4. The comparison is made by computing the error index known as mean square error (MSE).

3.3 Ant Colony/Bee Colony Optimization Testing

Our model succeeded in reducing tenth-order TFs to second-order TFs with average time domain error less than 0.8%, in a reasonable time. We simulated 50 TFs to verify our model. In this section, we introduce samples for our results, and we give some comparisons with other conventional MOR famous methods like moment-matching-based Padé approximation, Routh approximation, and balanced truncation method and prove how our method distinguishes among them in accuracy of frequency domain response, and time domain step response.

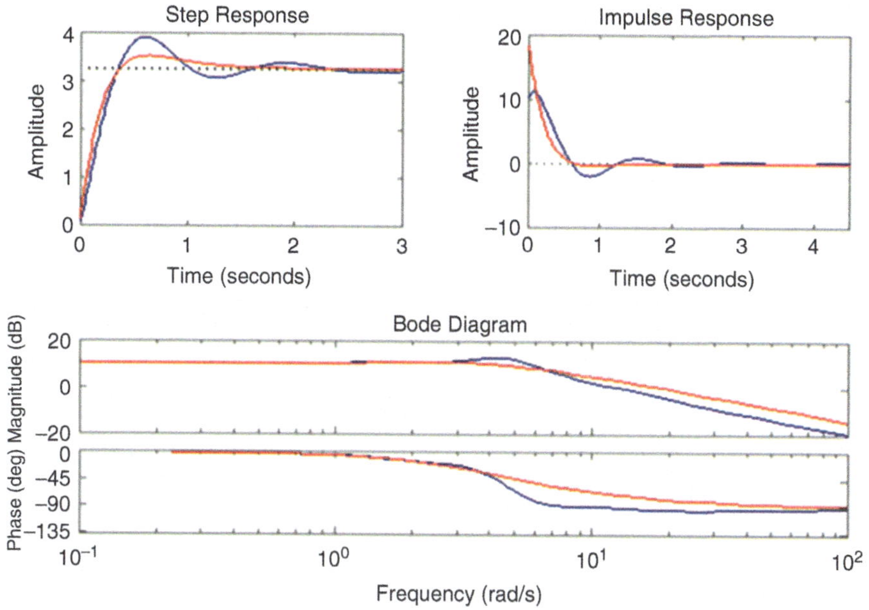

Fig. 4.4 Step response and frequency response comparison between original and reduced transfer function (Example 2)

4 Conclusions

This chapter presents a novel approach for MOR based on PSO, ANO, and ABC. Our approach aims at obtaining a reduced order model out of a relatively complex model represented as a TF. This ROM is obtained in a reasonable time. Moreover, it has accepted error in time domain presented by the step response, and the least error in frequency domain presented by Bode plot compared to other famous conventional methods in MOR like moment matching based Padé approximation, Routh algorithm, or balanced truncation.

References

1. C. Lanczos, An iteration method for the solution of the eigenvalue problem of linear differential and integral operators. J. Res. Natl. Bur. Stand. **45**, 225–280 (1950)
2. J. Kennedy and R.C. Eberhart, Particle Swarm Optimization, in *Proceedings of IEEE Conference on Neural Networks IV*, Piscataway, NJ, 1995
3. http://mnemstudio.org/particle-swarm-introduction.htm
4. Reduced Order Modeling of Linear MIMO Systems Using Particle Swarm Optimization
5. http://www.swarmintelligence.org/tutorials.php
6. S. Panda, S.K. Tomar, R. Prasad, C. Ardil, Reduction of linear time-invariant systems using Routh-approximation and PSO. Int. J. Appl. Math. Comput. Sci. **5**(2), 82–89 (2009)

7. A.K. Jagadev, S. Devi, R. Mall, Soft computing for feature selection, in *Knowledge Mining Using Intelligent Agents*, ed. by S. Dehuri, S. B. Cho (Imperial College Press, London, 2011), pp. 217–258
8. R.S. Parpinelli, H.S. Lopes, A.A. Freitas, An ant colony algorithm for classification rule discovery, in *Data Mining: A Heuristic Approach*, ed. by H. Abbass, R. Sarker, C. Newton (Idea Group Publishing, London, 2002), pp. 191–208
9. Z. Xin and F. Ping, An Improved Ant Colony Algorithm, in *Proceeedings of International Conference on MultiMedia and Information Technology*, 2008
10. D. Karaboga, An Idea Based on Honey Bee Swarm for Numerical Optimization, in *Technical report-TR 06, Erciyes University, Engineering Faculty, Computer Engineering Department*, 2005
11. D. Karaboga, B. Basturk, A powerful and efficient algorithm for numerical function optimization: Artificial bee colony (ABC) algorithm. J. Glob. Optim. **39**, 459–471 (2007)
12. D. Karaboga, B. Basturk, On the performance of artificial bee colony (ABC) algorithm. Appl. Soft Comput. **8**, 687–697 (2008)
13. D. Karaboga, B. Akay, A comparative study of artificial bee colony algorithm. Appl. Math. Comput. **214**, 108–132 (2009)
14. A. Sikander, R. Prasad, A novel order reduction method using cuckoo search algorithm. IETE J. Research **61**(2), 83–90 (2015)

Chapter 5
Control-Inspired Machine Learning Algorithm: Fuzzy Logic Optimization

1 Introduction

Fuzzy logic in simple words is a set of "if-then" rules. These rules describe the system behavior. These rules can be changed according to the required output. Fuzzy logic is a computational paradigm that generalizes classical two-valued logic for reasoning under uncertainty. In order to achieve this, the notation of membership in a set needs to become a matter of degree. This is the essence of fuzzy sets. By doing this one accomplishes two things: ease of describing human knowledge involving vague concepts and enhanced ability to develop a cost-effective solution to real-world problem. Fuzzy logic is a kind of multi-valued logic, which is a model-less approach and is a clever disguise of the Probability Theory. The most famous types of the membership functions are the triangle, trapezoidal, and Gaussian membership function.

Data can be divided into two sets: classical sets and fuzzy sets. Classical sets are the ordinary sets we are familiar with where each set has a sharp boundary. For example, a range of temperature from 0 to 100 is divided into low, medium, and high sets. Because each set has a sharp boundary any value can completely exist in one set or doesn't exist in it but it is not possible for one value to partially exist in a certain set. In contrast, Fuzzy logic defines the sets with non-sharp boundaries where there is a possibility for values to exist in other sets with certain probability. Due to fuzzy sets (Membership functions), the data exists in each set doesn't take crisp or sharp logic value (0 or 1 and belong or doesn't belong), but it can take any value between 0 and 1. This is the concept or fuzziness or the infinite logic [1].

Operations can be done on classical sets between two universe of discourse, for example, students in certain department at college (d_1, d_2, d_3, d_4) and students under 25 year (a_1, a_2, a_3, a_4). Those operations can be intersection, union, or difference. A intersect B is defined as the elements that exists in set A and in set B at the same time, A union B is defined as the values in both sets, and finally A difference B is defined as the elements in A without elements in B.

K. S. Mohamed, *Machine Learning for Model Order Reduction*,
https://doi.org/10.1007/978-3-319-75714-8_5

Fuzzy sets define the same operation, but with different implementation. First, since the elements inside a single universe of discourse can have varying degree between sets we can make operations between those sets, and second, for example, union is defined as the maximum of the two sets and intersection is defined as the minimum of the two sets. Simply both can be understood as OR and AND operations [2–5].

As previously stated, membership functions are the shape of the function defining the set. Many shapes can be used according to each application. Shapes include Triangular, Trapezoid, Bell, and Gaussian shape. Some of these shapes are simpler to implement like triangular and trapezoid, but have high dynamic variation that may be needed in certain applications, others like bell and Gaussian are hard to implement but they offer higher accuracy and they are more near to the human way of thinking.

Any fuzzy logic processing consists of three steps: Fuzzification, Interference, and Defuzzification. The input to any system is crisp. For example, if it is a temperature it can be 20, 23, 25, and so on. We need to map those values to our membership functions. The second step is interference where the output is concluded through set of operations and finally the output which is in fuzzy logic is passed through the defuzzification step where a crisp output suitable as a system output is deduced.

2 The Proposed Technique

2.1 Problem Formulation

Consider we have a linear system of nth order of q inputs and r outputs described using state space in time domain as follows:

$$H = Ax(t) + Bu(t) \tag{5.1}$$

$$Y(t) = Cx(t) \tag{5.2}$$

where H is n-dimensional state vector, u is q-dimensional control vector, and Y is r-dimensional output vector with $q \leq n$ and $r \leq n$. Also, A is $n \times n$ system matrix, B is $n \times q$ input matrix, and C is $r \times n$ output matrix.

By converting the state space model from time domain to frequency domain, the transfer function can be described in Eq. (5.3):

$$G(s) = \frac{N(s)}{D(s)} = \frac{\sum_{i=0}^{n-1} A_i s^i}{\sum_{i=0}^{n} a_i s^i} \tag{5.3}$$

where $N(s)$ is the numerator and $D(s)$ is the denominator of the higher order system. Also, A_i and a_i are the coefficients of numerator and denominator, respectively.

The problem is to find mth lower order model $R^m(s)$, where $m < n$ in the following form:

$$R(s) = \frac{N^m(s)}{D^m(s)} = \frac{\sum_{i=0}^{n-1} B_i s^i}{\sum_{i=0}^{n} b_i s^i} \tag{5.4}$$

In addition, we must take into consideration that the reduced model maintains the characteristics of the original transfer function.

2.2 *Fuzzy Logic Technique*

Fuzzy logic belongs to the family of many-valued logic. It focuses on fixed and approximate reasoning opposed to fixed and exact reasoning. A variable in fuzzy logic can take a truth value range between 0 and 1, as opposed to taking true or false in traditional binary sets. Since the truth value is a range, it can handle partial truth. Beginning of fuzzy logic was marked in 1956, with the introduction of fuzzy set theory by Lotfi Zadeh. Fuzzy logic provides a method to make definite decisions based on imprecise and ambiguous input data. Fuzzy logic is widely used for applications in control systems, since it closely resembles how a human make decision but in faster way.

The input to the fuzzy logic unit is two poles P_1, P_2 and the two numerators and the output with a single pole P. In case of complex conjugate pairs, 8-input to 2-output unit is used.

First, the membership functions and the rule are implemented. Many membership functions are used, but the most optimum approach found is the *Gaussian* membership functions which show lower dynamic variation and more accuracy. Spacing between membership function, their numbers, and intersection value determines accuracy of the tool and its range of operation. If the membership function covers poles values from 0 to 4, then the maximum value for pole is 4 and the least value is 0 and same thing happens for the constants.

Each unit reduces an order of two to order of one. If a transfer function of higher order is needed to be reduced, the operation is repeated several times. For example, an order 8 to 2 function reduction will reduce the 8 poles into 4 poles then the 4 new poles into the final 2 poles.

The whole reduction operation can be shown in the flow chart of Fig. 5.1. Unlike neural network, fuzzy logic does not come with a learning algorithm. The learning and identification of Fuzzy Models adopts techniques from other areas such as Statistics and Linear System Identification. The algorithm can be described by the following steps:

Step 1. Fuzzification: converts crisp inputs into a set of membership values in the interval [0, 1]. Example of Fuzzification step can be shown in Table 5.1.

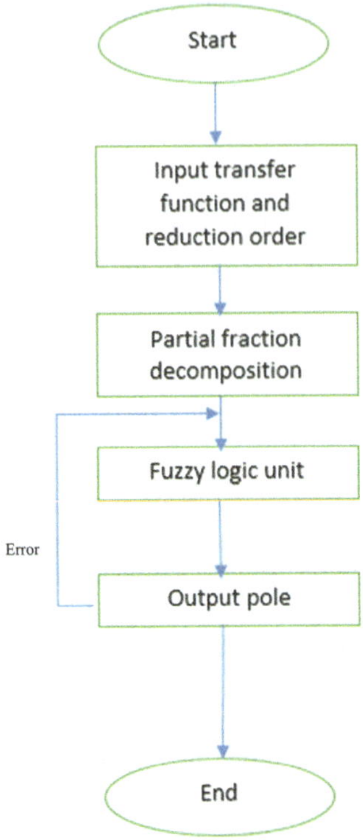

Fig. 5.1 Model order reduction using fuzzy logic flow chart

Table 5.1 Fuzzy rules example

No	Crisp input range	Fuzzy variable name
1	0–10	SMALL
2	10–100	AVERAGE
3	100–1000	LARGE

Table 5.2 Fuzzy rules example

Rule No	If	Then
1	Condition 1	Action 1
2	Condition 2	Action 2
3	Condition 3	Action 3

Step 2. Rule Inference: create rules to decide what action should be taken in response to the given set of degree of membership function. Example of rule inference can be shown in Table 5.2.

Step 3. Defuzzification: multiply fuzzy output obtained from the rules evaluation with its corresponding singleton value, then sum of this value is divided by the sum of all fuzzy output obtained from the rules evaluation. The result from this step is the final single output. This method is called the weighted average method. The defuzzification technique can be described by the following equation:

$$Defuz = \frac{\sum_{i=1}^{n} W_i P_i}{\sum_{i=1}^{n} W_i} \tag{5.5}$$

where P_i is the peak value of ith output membership function, W_i is the weight associated with ith rule, and n is the number of rules.

3 Testing of the Proposed Techniques

3.1 Setup of the Environment

The simulations are carried on Matlab 2012b software, core i7 processor, 6 GB Ram memory. The reduced order model is compared to the original one in terms of mean square error.

The testing procedures include a plot of step response for the reduced and the original TF, frequency response using Bode plot, mean square error, and the time elapsed.

The time taken for most of the original transfer functions to be reduced to lower order is nearly the same in all transfer functions, but sometimes some functions need more time than the others. The testing procedures are carried on more than 30 different transfer functions. The transfer functions have different forms of step responses and frequency responses where the step response of some of them has an exponential form and other responses were sinusoidal.

3.2 Fuzzy Logic Testing

The first example is transfer function with complex poles which produces sinusoidal response as depicted in Fig. 5.2. The error produced is 2.85% and time elapsed is 3.792 s. The second example is another transfer function with complex poles as shown in Fig. 5.3. The error produced is 5.65% and time elapsed is 2.752 s.

As can be seen from the results, the fuzzy logic introduces a whole new approach and thinking way toward model order reduction, but the algorithm is still not mature when used in this field. The main limitation of the algorithm is that there is restriction on the maximum value of pole which is fed to the fuzzy logic unit and that's because the nature of fuzzy logic work on bounded values covered by its membership function. If wider range of poles is to be used, the range of membership function must be increased.

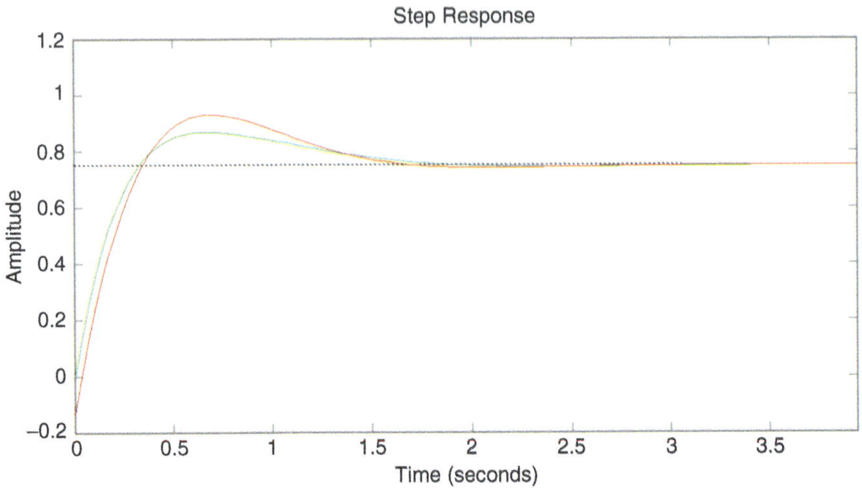

Fig. 5.2 Original model versus reduced one using fuzzy logic for example 1

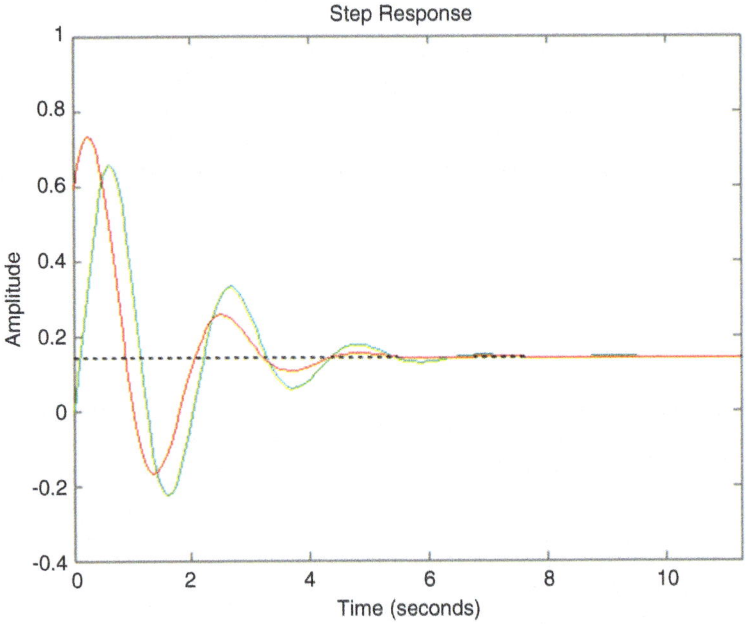

Fig. 5.3 Original model versus reduced one using fuzzy logic for example 2

4 Conclusions

This chapter presents a novel approach for MOR based on fuzzy logic. Our approach aims at obtaining a reduced order model out of a relatively complex model represented as a TF. This ROM is obtained in a reasonable time. Moreover, it has accepted error in time domain presented by the step response, and the least error in frequency domain presented by Bode plot compared to other famous conventional methods in MOR like moment matching based Padé approximation, Routh algorithm, or balanced truncation.

References

1. S.N. Sivanandam, *Introduction to Fuzzy Logic Using Matlab* (Springer, Berlin, 2012)
2. M.M. Algazar, H. AL-monier, H.A. EL-halim, M.E. El Kotb Salem, Maximum power point tracking using fuzzy logic control. Int. J. Electr. Power Energy Syst. **39**(1), 21–28 (2012)
3. G. Anil, Fuzzy logic based maximum power point tracker for a PV system. IOSR J. Electr. Electron. Eng. **54**, 13–21 (2013), e-issn: 2278-1676, p-ISSN: 2320-3331
4. J. Alvarez, A. Lago, A. Nogueiras, FPGA implementation of a fuzzy controller for automobile DC-DC converters, in *IEEE International Conference on Field Programmable Technology*, Dec. 2006, pp. 237–240, ISBN: 0-7803-9729-0
5. M.S. Alam, M.F. Azeem, A.T. Alouani, Modified Queen-Bee algorithm-based fuzzy logic control for real-time robust load matching for a solar PV system. IEEE Trans. Sustainable Energy **5**(2), 691–698 (2013)

Chapter 6
Brain-Inspired Machine Learning Algorithm: Neural Network Optimization

1 Introduction

An artificial neural network (ANN) is a network inspired by biological neural networks (the central nervous systems of animals, in particular the brain) which are used to estimate or approximate functions that can depend on a large number of inputs that are generally unknown [1–7]. All inputs for the given neuron are multiplied by their respective weights, summed up, and then transformed using an activation function. Activation functions usually restrict the range of values of the neuron output, for example, from minus one to one. This transformed value then serves as the input for the next neuron. A neural network consists of three types of layers—input, hidden, and output as depicted in Fig. 6.1. The input layer contains the data supplied to the network. In the hidden layers the transformation of the inputs takes place and the output layer produces the estimate. The number of hidden layers is not limited. Several activation functions exist and they also influence the performance of the ANNs. Perhaps the simplest activation function is threshold function. If the signal from a neuron is larger or equal to a certain set value, then the output value is equal to 1 else the output value is equal to zero. This type of signal is good for binary problems. In general, ANNs learn from the data and they are able to improve in iterations by adjusting the weights [7].

2 The Proposed Technique

2.1 Problem Formulation

Consider we have a linear system of nth order of q inputs and r outputs described using state space in time domain as follows:

K. S. Mohamed, *Machine Learning for Model Order Reduction*,
https://doi.org/10.1007/978-3-319-75714-8_6

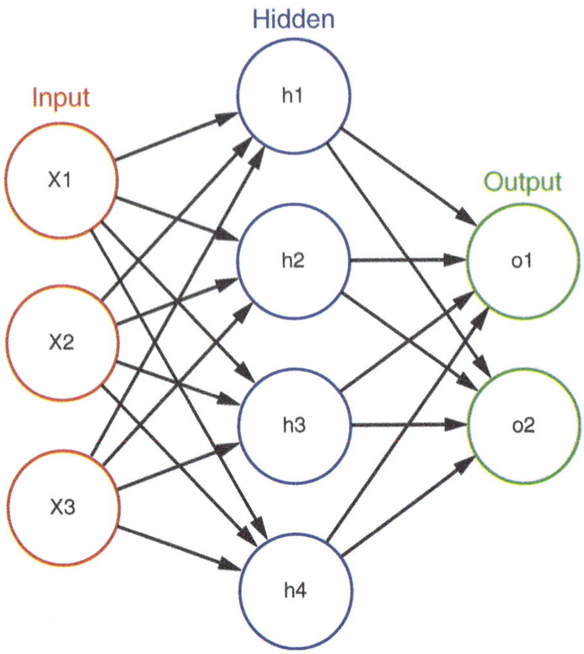

Fig. 6.1 The architecture/topology of the feed-forward ANN layers. The neurons are organized in the form of layers. The neurons in a layer get inputs from the previous layer and feed their output to the next layer. Output connections from a neuron to the same or previous layer neurons are not permitted. The maximum number of parallelism equals the number of neurons. The layers of neural network are: Input Layer, Hidden Layer, and Output Layer

$$H = Ax(t) + Bu(t) \tag{6.1}$$

$$Y(t) = Cx(t) \tag{6.2}$$

where H is n-dimensional state vector, u is q-dimensional control vector, and Y is r-dimensional output vector with $q \leq n$ and $r \leq n$. Also, A is $n \times n$ system matrix, B is $n \times q$ input matrix, and C is $r \times n$ output matrix.

By converting the state space model from time domain to frequency domain, the transfer function can be described in Eq. (6.3):

$$G(s) = \frac{N(s)}{D(s)} = \frac{\sum_{i=0}^{n-1} A_i s^i}{\sum_{i=0}^{n} a_i s^i} \tag{6.3}$$

where $N(s)$ is the numerator and $D(s)$ is the denominator of the higher order system. Also, A_i and a_i are the coefficients of numerator and denominator, respectively.

The problem is to find a mth lower order model $R^m(s)$, where $m < n$ in the following form:

$$R(s) = \frac{N^m(s)}{D^m(s)} = \frac{\sum_{i=0}^{n-1} B_i s^i}{\sum_{i=0}^{n} b_i s^i} \tag{6.4}$$

In addition, we must take into consideration that the reduced model maintains the characteristics of the original transfer function.

2.2 Neural Network Technique

Neural Networks are computational models that consist of nodes that are connected by links. Each node performs a simple operation to compute its output from its input, which is transmitted through links connected to other nodes. This relatively simple computational model is Artificial Neural Network because the structure is analogous to that of Neural Systems in Human Brains—nodes corresponding to neurons and links corresponding to synapses that transmit signal between neurons. One of the major features of a Neural Network is its learning capability. While the details of learning algorithms of Neural Networks vary from architecture to architecture, they have one thing in common; they can adjust the parameters in a Neural Network such that the network learns to improve its performance of a given task.

The ANN used in this work is a feed-forward neural network which has the advantage of being applicable to static or dynamic system matrices (adaptive modeling).

In addition to that, there is a one-to-one correspondence between the ANN weights and the system matrix elements. The network consists of a one-neuron-layer network with linear neuron activation function along with the given input and output data.

Using ANN, the discrete full-order system will be reduced to a proper-order model. The reduced model is directly obtained based on the cost function minimization.

To start with, the output matrix is designed such that $C_r = [1\ 0\ 0\ \ldots\ldots\ 0]$ which, in some way, is similar to the controllable canonical form where the output is one of the dynamical states. In order to determine the reduced order system matrices, the neural network training is implemented.

The general architecture of the ANN is shown in Fig. 6.2. Common activation filtering function is the sigmoid function. The input to the activation function is given by:

$$a = \sum\nolimits_{j=1}^{N} w_j x_j \tag{6.5}$$

The output from the activation function is given by:

$$y = f(a) \tag{6.6}$$

The sigmoid activation function is given by:

$$y = \frac{1}{1 + e^{-x}} \tag{6.7}$$

The sigmoid activation function can be approximated by piecewise linear approximation (PWL) to have five linear segments as shown in Eq. (6.4) [8]. The corresponding block diagram of the approximated sigmoid function is shown in Fig. 6.2. The main aim of the activation function is to compress an infinite input

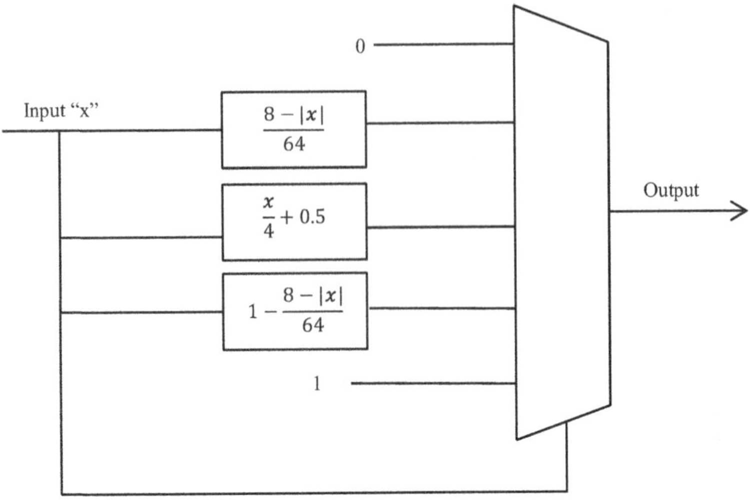

Fig. 6.2 The block diagram of the sigmoid activation function

Fig. 6.3 The sigmoid function

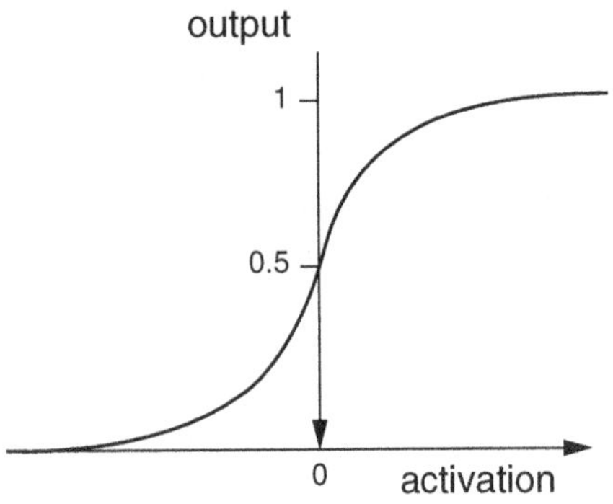

range to a finite output range, e.g., [−1, +1]. The bounded sigmoid function is shown in Fig. 6.3.

$$f(x)=\begin{cases} 0, & x\le -8 \\ \dfrac{8-|x|}{64}, & -8<x\le -1.6 \\ \dfrac{x}{4}+0.5, & |x|<1.6 \\ 1-\dfrac{8-|x|}{64}, & 1.6\le x\le 8 \\ 1, & x>8 \end{cases} \tag{6.8}$$

Neural networks can be trained to solve problems that are difficult to be solved by conventional algorithms. Training refers to an adjustment of the connection weights, based on a number of training examples that consist of specified inputs and

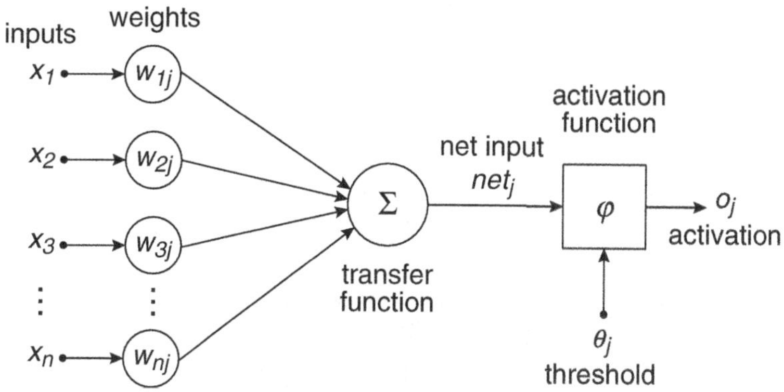

Fig. 6.4 The general architecture of ANN. The artificial neuron given in this figure has N inputs, denoted as $x_1, x_2, \ldots, x_N$. Each line connecting these inputs to the neuron is assigned a weight, denoted as $w_1, w_2, \ldots, w_N$, respectively. The activation, a, determines whether the neuron is to be fired or not. The activation function used is sigmoid function

corresponding target outputs. Training is an incremental process where after each presentation of a training example, the weights are adjusted to reduce the discrepancy between the network and the target output. Popular learning algorithms are variants of gradient descent (e.g., error-backpropagation), and radial basis function adjustments. The back propagation algorithm is described in Fig. 6.3, where it is a supervised learning algorithm. The input and output are known prior to the computation. The signals from the input layer propagate to the output layer through the hidden neurons in a feed-forward manner, the output is compared with the expected value and the difference in the error is back propagated making the weights to be updated. There are basically two ways of propagation: a forward path and a backward path. In forward path the weights are fixed and signals moves in forward direction whereas in backward path the weights are adjusted according to the calculated error. The difference in the actual and obtained values of output are measured and the error is propagated backwards (Figs. 6.4 and 6.5).

3 Testing of the Proposed Techniques

3.1 Setup of the Environment

The simulations are carried on Matlab 2012b software, core i7 processor, 6 GB Ram memory. The reduced order model is compared to the original one in terms of mean square error.

The testing procedures include a plot of step response for the reduced and the original TF, frequency response using Bode plot, mean square error, and the time elapsed.

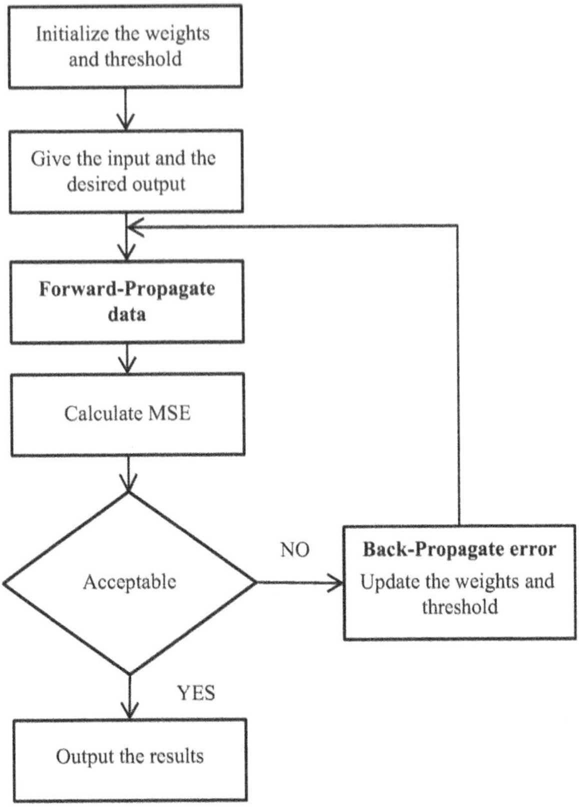

Fig. 6.5 The flow-chart of back propagation algorithm

The time taken for most of the original transfer functions to be reduced to lower order is nearly the same in all transfer functions, but sometimes some functions need more time than the others. The testing procedures are carried on more than 30 different transfer functions. The transfer functions have different forms of step responses and frequency responses where the step response of some of them has an exponential form and other responses were sinusoidal.

3.2 Neural Network Testing

The results of reducing the original model of a 10th-order transfer function to a second-order model is analyzed where the step and frequency response are shown in Figs. 6.1 and 6.2, respectively.

A comparison has been made between the results of the proposed model and the results of mathematical methods of Padé approximation, Routh approximation, and truncated-balanced method. The proposed model shows efficiency in terms of

Table 6.1 Shows a comparison of the results of all methods for the first transfer function

Metric	ANN	Padé	Routh	Balanced
MSE %	0.14	1760	5.28	0.05
Time (s)	4	0.1	0.1	0.1

Table 6.2 Example #1

Original order	6
Desired order	2
Original TF	Num: $0.5 \times 10^3 \times$ [0.0010 0.0275 0.2977 1.5722 4.0561 4.1137] Den: $10^3 \times$ [0.0010 0.0241 0.2120 0.8382 1.5052 1.0706 0.2559]
Reduced TF	Num = [0.9808 3.7437]; Den = [1 1.6587 0.4582];
Mean square error	2.35%
Run time	23 s

Table 6.3 Example #2

Original order	7
Desired order	2
Original TF	Num = 1.0e+06 * [0 0.0000 0.0001 0.0014 0.0178 0.1337 0.5781 1.3280 1.2502]
	Den = 1.0e+05 * [0.0000 0.0004 0.0060 0.0587 0.3699 1.5866 4.6564 8.5433 7.5835]
Reduced TF	[2.4834 23.4423]
	[1 2.2530 14.0569]
Mean square error	3.1%
Run time	19 s

accuracy over Routh approximation while the results were nearly equal compared with Padé approximation but the accuracy of Padé approximation was bad and sometimes unacceptable when testing transfer function of sinusoidal step response. Truncated-balanced method showed efficiency concerning accuracy over the proposed model but the proposed model showed efficiency over it concerning frequency response as the frequency response of the Truncated-balanced method is not accurate enough in most of the test cases and sometimes completely different from the original model. Ready schemes for the mathematical methods used in the experiments. The results of the proposed model and the mathematical methods are compared in Table 6.1. More examples are shown in Tables 6.2, 6.3, 6.4, and 6.5 (Figs. 6.6 and 6.7).

Table 6.4 Example #3

Original order	8
Desired order	2
Original TF	Num: [0 1 48.6902883638322 1061.36522193613 13036.0107753892 95837.1206975369 412845.211754671 956623.548032344 918252.595867870]
	Den: [1 25.6936779696381 329.665759427945 2535.24673612937 14138.1393057336 49575.0063353493 135879.971107481 173685.834110139 251086.931791740]
Reduced TF	[8.7219 14.3283]
	[1 0.7868 3.8815]
Mean square error	2.1%
Run time	25 s

Table 6.5 Example#4

Original order	10
Desired order	2
Original TF	Num: [1 39.2474195915156 684.147508905409 7013.28745078928 46853.9667412056 212386.706779798 655120.949927384 1328964.94877038 1611930.84460820 892547.414599374]
	Den: [1 51.8283488196718 1225.57888810214 17379.9035520574 163748.484352740 1074179.89831544 4998071.67285878 16438852.9258060 37130762.1640610 53626302.0877531 40227301.0613871]
Reduced TF	[1.0277 1.2740]
	[1 15.0456 59.1541]
Mean square error	2.4%
Run time	28 s

4 Conclusions

This chapter presents a novel approach for MOR based on artificial neural network. Our approach aims at obtaining a reduced order model out of a relatively complex model represented as a TF. This ROM is obtained in a reasonable time. Moreover, it has accepted error in time domain presented by the step response, and the least error in frequency domain presented by Bode plot compared to other famous conventional methods in MOR like moment matching based Padé approximation, Routh algorithm, or balanced truncation. Artificial neural network is an effective method in model order reduction.

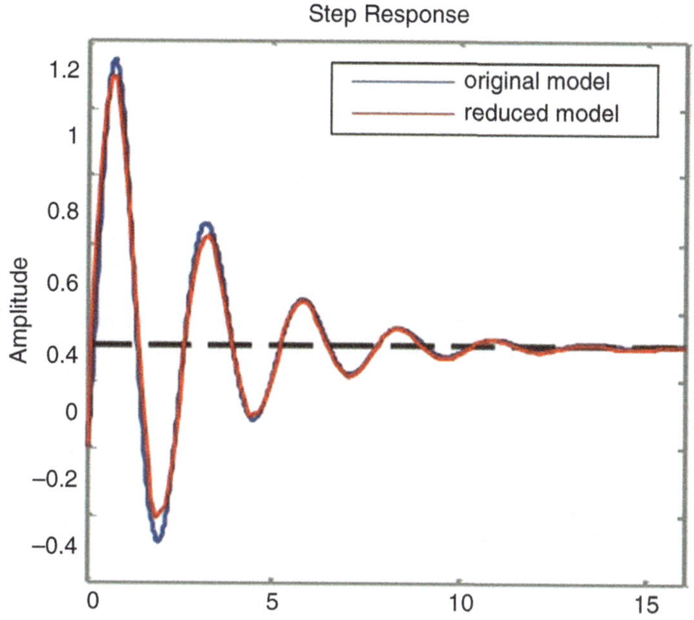

Fig. 6.6 The step response in case of ANN

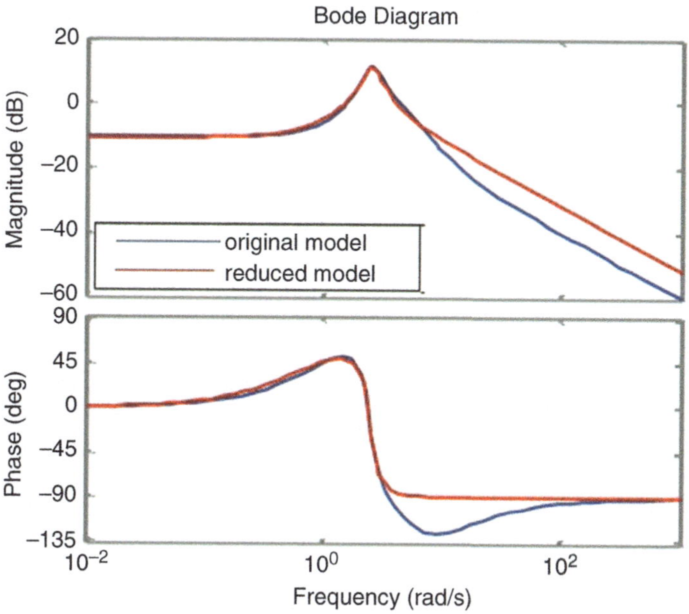

Fig. 6.7 The frequency response of the original and reduced transfer models for G1(s) with mag. MSE 0.77%

References

1. S. Wilhelmus, A. Henk, V.D. Vorst, R. Joost, *Model Order Reduction: Theory, Research Aspects and Applications* (Springer, Berlin, 2008)
2. A.C. Antoulas, Approximation of Large-Scale Dynamical Systems, in *Advance in Design and Control*, (SIAM, Philadelphia, 2005)
3. F.H. Bellamine, Model order reduction using neural network principal component analysis and generalized dimensional analysis. Int. J. Comput. Aided Eng. Softw. **25**(5), 443–463 (2008)
4. M.A. Christodoulo, T.N. Liopoulos, Neural network models for prediction of steady state and dynamic behavior of MAPK cascade, in *Proceedings of the 14th Mediterranean Conference on Control and Automation*, Ancona, June 28–30, 2006, pp. 1–9
5. Y. Chetouni, Model-order reduction based on artificial neural networks for accurate prediction of the product quality in a distillation column. Int. J. Autom. Control **3**(4), 332–351 (2009)
6. S. Haykin, *Neural Networks: A Comprehensive Foundation* (Macmillan College Publishing Company, New York, 1994)
7. J.G. Kusckewschi, S. Zein-Sabatto, Discrete dynamic systems parameter identification and state estimation using a recurrent neural network, in *World Congress on Neural Network (WCNN'95) Conference*, Washington DC, July 1995, pp. 92–95
8. J. Holt, T. Baker, Back propagation simulations using limited precision calculations, in *Proceedings of International Joint Conference on Neural Networks (IJCNN-91)*, Seattle, WA, July 1991, vol. 2, pp. 121–126

Chapter 7
Comparisons, Hybrid Solutions, Hardware Architectures, and New Directions

1 Introduction

In this chapter, the performance of five machine learning algorithms such as GA (Genetic Algorithm), ANN (Artificial Neural Network), PSO (Particle Swarm Optimization), FL (Fuzzy Logic), and SA (Simulated Annealing) for the problem model order reduction of linear systems are compared. In addition, hybrid solutions are presented. Finally, new directions in machine learning are discussed.

2 Comparisons Between the Different Algorithms

In Table 7.1, we present the final comparison among both the five machine-learning-based techniques and the conventional methods. In the following table, we introduce samples of comparisons between our algorithms and the other conventional MOR algorithms. The numerical comparisons prove the advantage of our algorithm on the other algorithms in terms of frequency response. Moreover, our algorithms are iterative algorithm, which means that the more iterations the more accuracy we obtain, until they reach the most satisfying result to the user. This feature distinguishes our algorithms and makes them application-dependent. If the user wants higher accuracy, he should increase the number of iterations, while if his application is dependent on the speed of the output, the user should decrease the number of iterations that conserves the acceptable percentage of error, which is nearly (<5%). Neural Networks has proven to be more efficient if the focus is the processing speed. The control parameter values for all the optimization algorithms are given in Table 7.2.

K. S. Mohamed, *Machine Learning for Model Order Reduction*,
https://doi.org/10.1007/978-3-319-75714-8_7

Table 7.1 The final comparison among the five proposed techniques and the conventional methods

	SA	GA	ANN	PS	FL	Padé
MSE %	1.5	2.16	1.7	2.4	1.9	15.36
Run-time (s)	80	23	4	56	33	0.1

Table 7.2 The control parameter values for all the optimization algorithms

	SA	GA	ANN	PSO	FL
Population	30	30	30	30	30
Generation	300	300	300	300	300
Crossover probability	–	0.5	–	–	–
Mutation probability	–	0.1	–	–	–
Number of neighbors	–	–	–	5	–
Social learning factor	–	–	–	0.5	–
Cognitive learning factor	–	–	–	2	–
Cooperative factor	–	–	–	2	–
Initial cooling temperature	100	–	–	–	–
Cooling constant	0.9	–	–	–	–
Number of hidden layers	–	–	1	–	–
Learning rate	–	–	0.2	–	–
Number of rules	–	–	–	–	7

3 Hybrid Solutions Between the Different Algorithms

The main disadvantage of the optimization algorithms such as SA and GA is that they depend on the number of iterations. Accordingly, a new combined simulated annealing (SA) and genetic algorithm (GA) approach is proposed to solve the problem of Model Order Reduction (MOR). This novel method involves two steps: a set of solutions is generated first and then the best subset of these solutions is selected by the combined SA and GA algorithm [1–6].

SA is the main process in the procedures to search for a better solution to minimize the cost function (MSE) in our problem. Besides, GA is used as a subprocess to generate new solutions.

Several GA operators (crossover and mutation) are utilized and tested for their effectiveness. The results show that the proposed method can efficiently converge to the optimal solution on different problems.

With the increase of computer power, novel approaches emerged for solving complex optimization problems with large search spaces. Genetic algorithm (GA) is one of the popular methods and has been applied to optimization problems. However, it is known that GA may stick in a local optimum as it only accepts better solutions in its reproduction process. Yet accepting a worse solution is sometimes beneficial as it may lead to searching a wider space and finally reaching the global optimum [7–10].

Simulated annealing (SA) is also a promising algorithm with a strong search capability. Although SA is not always capable of finding the global optimal solution, SA can sometimes accept worse solutions when certain requirements are met so as to extend its searching space beyond the local optimum.

The idea of combining those two powerful algorithms in one algorithm to take the advantages of those two algorithms inspired from our studies to those algorithms. It is promising to think that the combination of two powerful search algorithms may achieve better results, especially, when one algorithm complements the other as in. Consequently, a combined SA and GA approach is proposed in this work.

The Hybrid algorithm is implemented by applying genetic algorithm to the selected model to be reduced producing a solution, and then the simulated annealing is applied to the reduced model creating a better chance in finding the most optimum solution. The population in the genetic algorithm was set to 2 and the number of iterations was set to 100. In simulated annealing algorithm the number of iterations was set to 500 and the initial temperature was set to 1000 [11, 12].

The simulations are carried on Matlab 2012b software, core i7 processor, 6 GB Ram memory. The reduced order model is compared to the original one in terms of mean square error.

The testing procedures includes a plot of step response for the reduced and the original TF, frequency response using Bode plot, mean square error, and the time elapsed.

The time taken for most of the original transfer functions to be reduced to lower order is nearly the same in all transfer functions, but sometimes some functions need more time than the others. The testing procedure was carried on more than 30 different transfer functions. The transfer functions have different forms of step responses and frequency responses where the step response of some of them has an exponential form and other responses were sinusoidal. The proposed model showed a great efficiency in terms of accuracy and preserving the original model's properties. The results for some transfer functions testing are shown in Tables 7.3 and 7.4.

The results of our approach show that the proposed method succeeds to dramatically reduce the expensive computations of the GA algorithm, hence, the run-time of the code. The disadvantage of that method is that it sometimes gives low accuracy (high MSE) due to the lower number of iterations, but is still better than each algorithm independently.

Table 7.3 The results of the reduced order model for 6th-order TF

Metric	GA	SA	Hybrid
Time elapsed (s)	23	95	11
MSE %	2.5	3.1	4

Table 7.4 The results of the reduced order model for 8th-order TF

Metric	GA	SA	Hybrid
Time elapsed (s)	21	23	12
MSE %	0.5	1.7	2

Table 7.5 A comparison of the results of all methods for the first transfer function

Metric	Hybrid	Padé	Routh	Balanced
Step response MSE (%)	0.03	1760	5.28	0.05
Freq. response mag. MSE (%)	0.62	34	100	0.63
Time (s)	11	0.1	0.1	0.1

Table 7.6 A comparison of the results for all methods for the second transfer function

Metric	Hybrid	Padé	Routh	Balanced
Step response MSE (%)	0.04	1760	5.28	0.05
Freq. response mag. MSE (%)	0.36	0.33	23.8	0.19
Time (s)	12	0.1	0.1	0.1

A comparison has been made between the results of the proposed model and the results of mathematical methods of Padé approximation, Routh approximation, and truncated-balanced method. The proposed model showed efficiency in terms of accuracy over Routh approximation while the results were nearly equal compared with Padé approximation, but the accuracy of Padé approximation was bad and sometimes unacceptable when testing transfer function of sinusoidal step response.

Truncated-balanced method showed efficiency concerning accuracy over the proposed model, but the proposed model showed efficiency over it concerning frequency response as the frequency response of the truncated-balanced method was not accurate enough in most of the test cases and sometimes completely different from the original model. Ready schemes for the mathematical methods used in the experiments [12].

The results of the proposed model and the mathematical methods are compared in Table 7.5 and 7.6 for the two TFs.

The experiments compared the proposed model with powerful mathematical methods showing the efficiency of the proposed model over the mathematical model in terms of accuracy and preserving the properties of the original model. Hybrid-based MOR is a promising solution in reducing the circuit-level simulation time.

Moreover, sometimes fuzzy logic can be used with neural network as shown in Fig. 7.1. Or GA with PSO [14]. Hybrid GA and PSO are remarkably superior to stand-alone PSO and GA. Also, Hybrid GA-ANN is a promising MOR technique. The flowchart of the hybrid GA-ANN is shown in Fig. 7.2, where ANN is used to initialize the GA to reduce the computation time. Another possible combination is between GA and FL as proposed in [15]. In [16] , neural network has been combined with a simulated annealing algorithm to improve optimization problems. Moreover, PSO can be combined with ANN as proposed in [17]. A hybrid algorithm of particle swarm optimization (PSO) and simulated annealing (SA) algorithm is proposed in [18]. In [19], PSO combined with FL is proposed for optimization problems. SA is used with FL in [20]. All the possible hybrid solutions are summarized in Table 7.7.

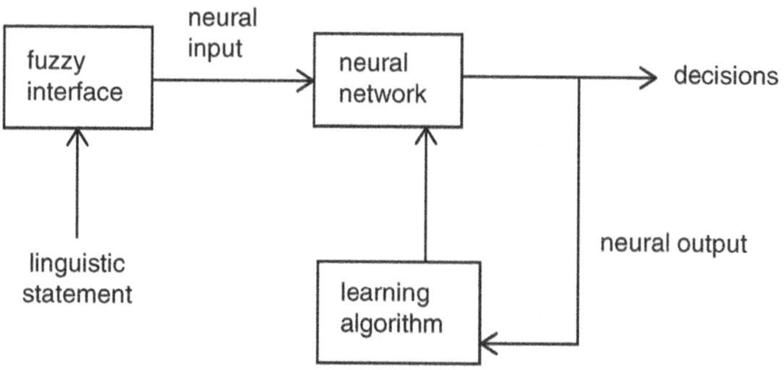

Fig. 7.1 Fuzzy logic with neural network

4 Hardware Architecture

The block diagram of the proposed GA algorithm is shown in Fig. 7.3.

The proposed ANN architecture is shown in Fig. 7.4. The neurons are organized in the form of layers. The neurons in a layer get inputs from the previous layer and feed their output to the next layer. Output connections from a neuron to the same or previous layer neurons are not permitted. The maximum number of parallelism equals the number of neurons. The data flow is such that the weights from the weights block and the inputs from outside are multiplied and the products are summed in the adder block and accumulated in the register, and then the sum of product is entering the sigmoid activation function block then forwarded to the cost function block. According to the calculated mean square error, the weights are updated using back propagation algorithm. Thee multiplier is 8x8 bit multiplier and the adder is 64 bit adder. The approximated sigmoid function block diagram is shown in Fig. 7.5.

The proposed fuzzy system architecture consists of three modules that are fuzzifier, rule inference and defuzzifier as depicted in Fig. 7.6.

The block diagram of the proposed PSO algorithm is shown in Fig. 7.7.

The block diagram of the proposed *simulated annealing* algorithm is shown in Fig. 7.8.

5 New Directions in Machine Learning

5.1 Chaotic Genetic Algorithm

We can use chaotic maps to generate chaotic values instead of the random values in GA processes. This can reduce significantly the number of iterations. The diversity of the Chaos Genetic Algorithm (CGA) avoids local convergence more often than the traditional GA. Chaos is a behavior in a dynamic system that means it is highly

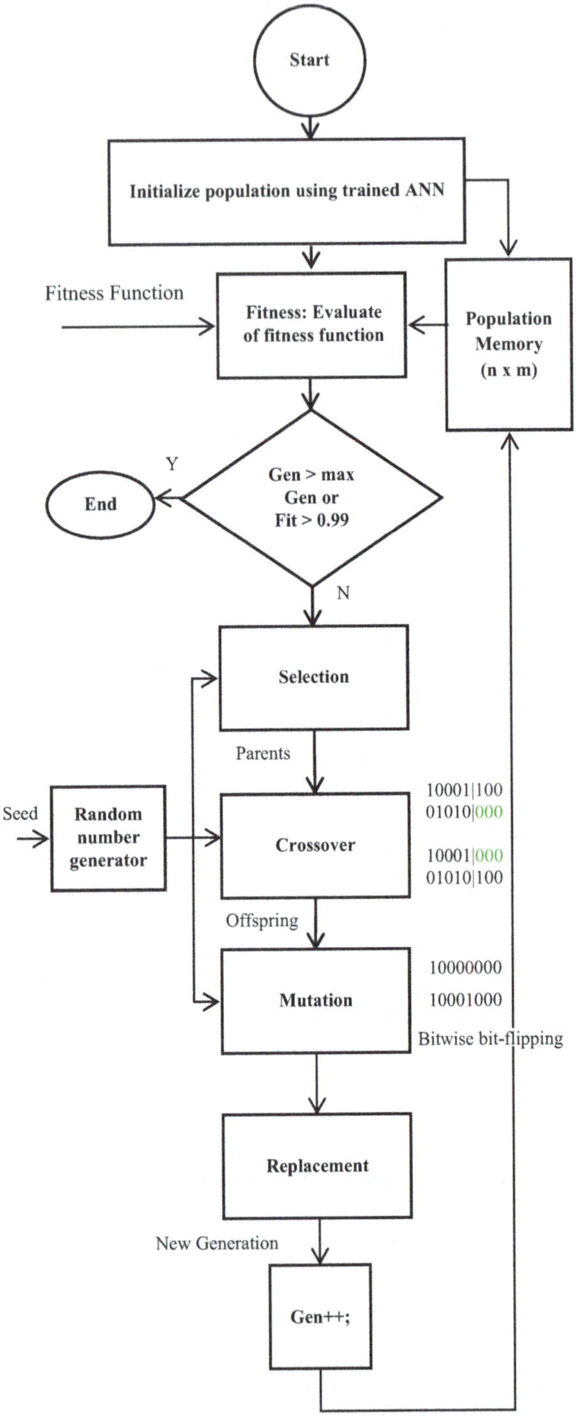

Fig. 7.2 Hybrid GA-ANN flow-chart. ANN is used to initialize the GA

Table 7.7 All the possible hybrid solutions

	SA	GA	ANN	PSO	FL
SA	✓	✓	✓	✓	✓
GA	✓	✓	✓	✓	✓
ANN	✓	✓	✓	✓	✓
PSO	✓	✓	✓	✓	✓
FL	✓	✓	✓	✓	✓

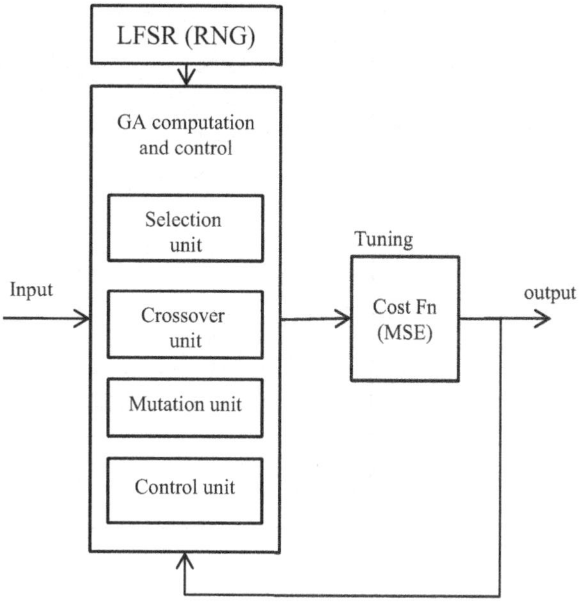

Fig. 7.3 The block diagram of the proposed GA algorithm. The pseudo random numbers are generated in hardware by using a linear feedback shift register (LFSR). The GA computing and controller block consist of selection, crossover, mutation, and control sub-modules

sensitive to changes of initial condition. A small change to initial condition can lead to a big change in the behavior of the system. The logistic maps can be used in the selection, crossover, and mutation phase [12].

5.2 *Spiking Neural Network Algorithm*

Spiking Neural Networks (SNNs) are widely regarded as the third generation of artificial neural networks. SNNs have great potential for improving the accuracy and efficiency of conventional neural networks through event-based computation. Spiking concept means the activation of individual neurons. Also, spikes mean pulses. There are many different schemes for the use of spike timing information in neural computation. In SNNs, information is represented and processed using spikes, which take a binary value 0 or 1. Inputs are presented to the neurons in the first layer as a time series of spikes. The spike trains propagate through the network until the output layer is reached. Each neuron in the output layer is associated with a class label, and

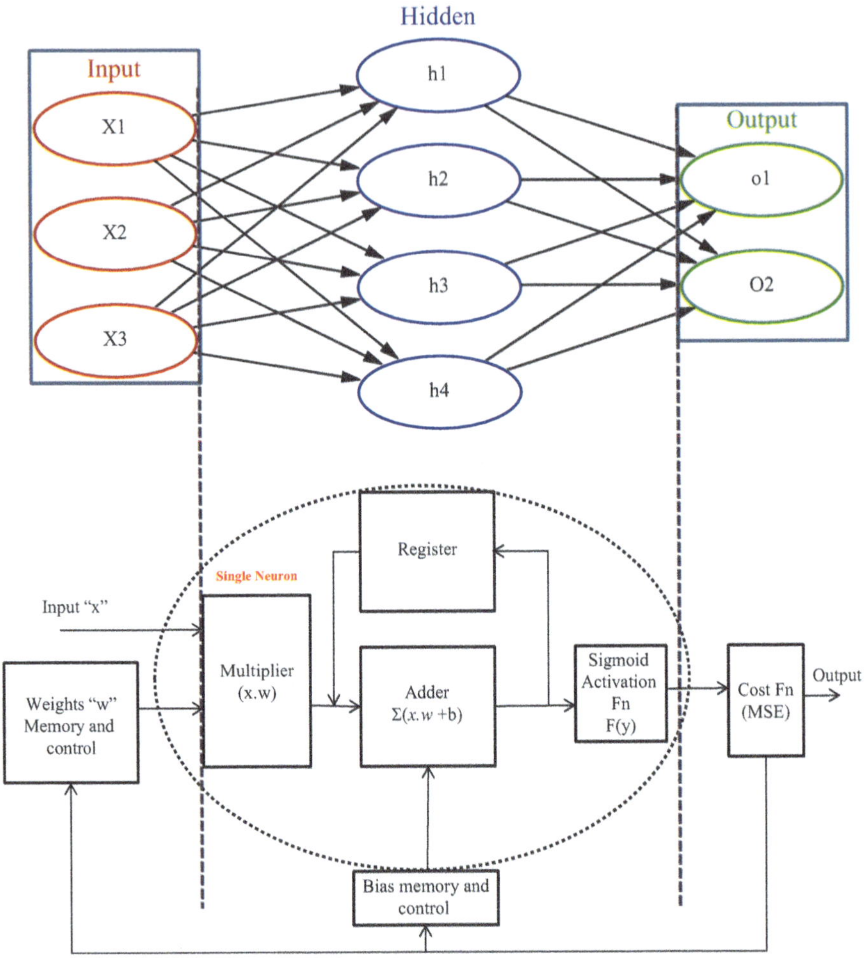

Fig. 7.4 The block diagram of ANN

the input is assigned the class corresponding to the output neuron that spiked the largest number of times. The number of time steps for which the SNN is evaluated is a key network parameter, and is determined during training [13].

5.3 Differential Algorithm

The differential method is based on differentiation of polynomials. The reciprocals of the Numerator and Denominator Polynomial of the higher order Transfer Functions are differentiated successively many times to yield the coefficients of the reduced order transfer function [21]. The reduced polynomials are reciprocated

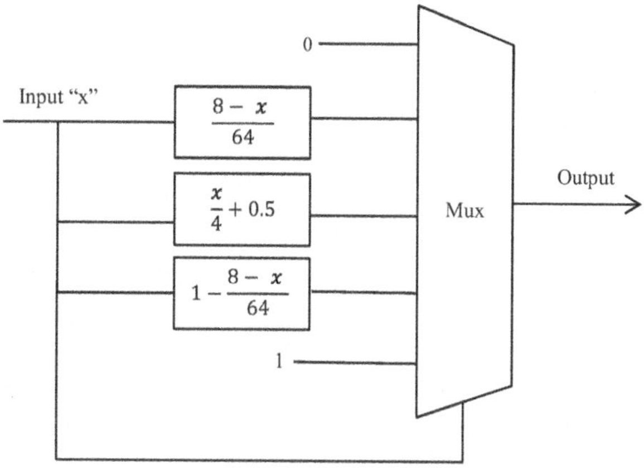

Fig. 7.5 The block diagram of the sigmoid activation function

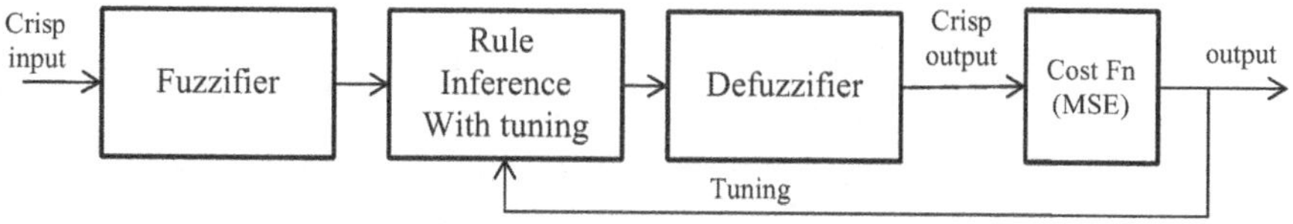

Fig. 7.6 The block diagram of the proposed fuzzy logic algorithm

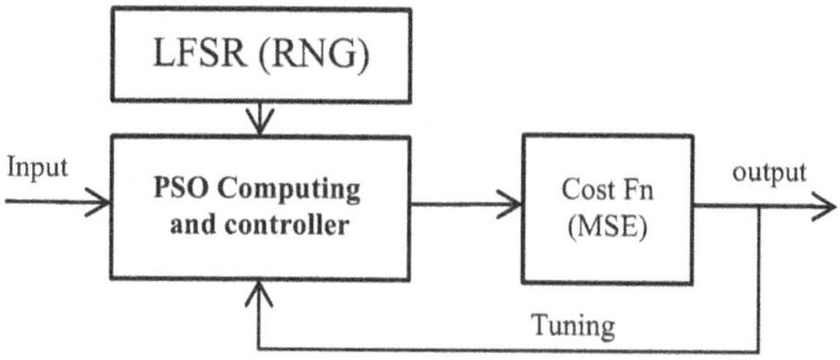

Fig. 7.7 The block diagram of the proposed PSO algorithm. The pseudo random numbers are generated in hardware by using a linear feedback shift register (LFSR). This is a technique based on a shift register design which uses several special bits named taps which are fed forward for generating a new number. Normally, xor is used as a feed-forward function. LFSRs are implemented for the initial values of weights, velocities, Pbests, and Gbest, and coefficients in the velocity update equation

back and normalized. The proposed method is described in Fig. 7.9 and it can be summarized in Listing 1.

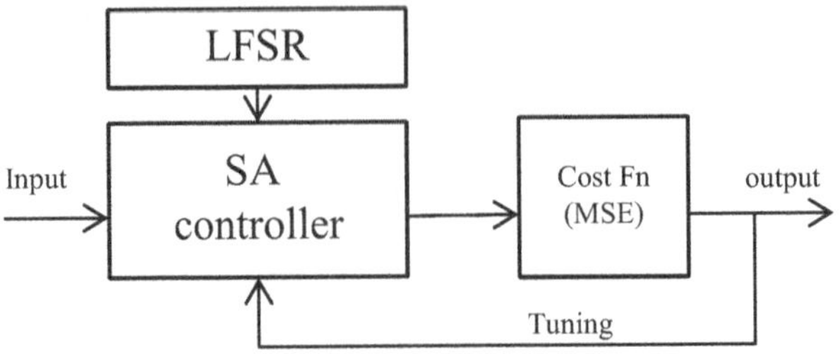

Fig. 7.8 The block diagram of the proposed simulated annealing algorithm

Fig. 7.9 The procedure to obtain reduced order model using proposed method

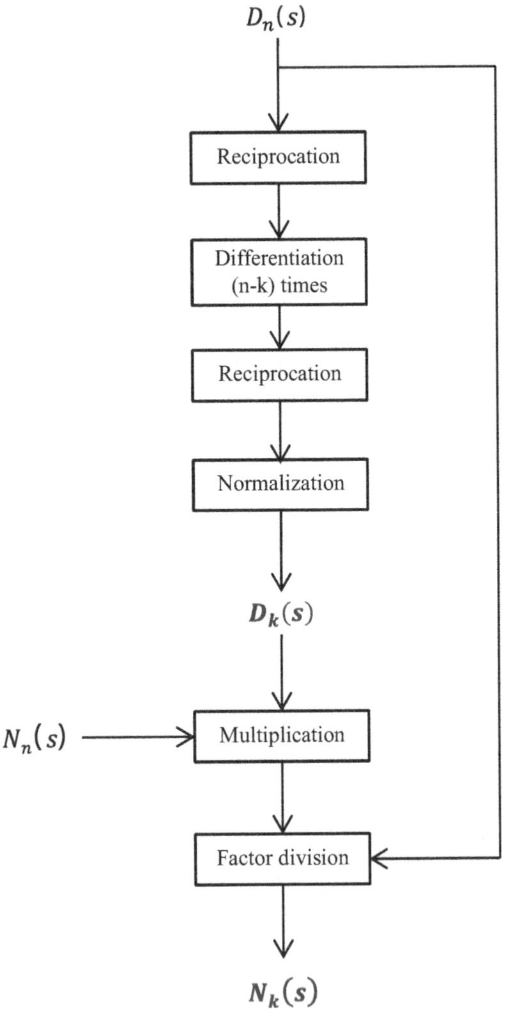

Listing 1 The Proposed Algorithm's Procedure
Procedure: Differentiation algorithm
1: **Input:** $N_n(s)$, $D_n(s)$, k
2: **Output:** $N_K(s)$, $D_K(s)$.
3: Get the reciprocation of $D(s)$.
4: Differentiate it (n-k) times.
5: Reciprocate the result.
6: Normalize it to get $D_K(s)$.
7: Multiplication of $N_n(s)$ *and* $D_K(s)$.
8: Factor division of the result by coefficient of $D_n(s)$

5.4 *Model Order Reduction Using Cross Multiplication of Polynomials*

Our purpose is to reduce the high order model described by Eq. (7.1) from the *n*th order to the *r*th order as shown in the TF shown in Eq. (7.2) where ($r < n$). The proposed method is using cross multiplication of the polynomials, where we cross multiply the two polynomials of the unreduced model and the reduced model. Multiplying (7.1) and (7.2) and comparing the coefficients of same power results in Eqs. (7.3) and (7.4). Using genetic algorithm can speedup the process of obtaining the coefficients of the model order reduction. Assuming the initial conditions given by Eq. (7.5) and (7.6).

$$H(S) = \frac{a_0 + a_1 S + a_2 S^2 + \cdots + a_{n-1} S^{n-1}}{b_0 + b_1 S + b_2 S^2 + \cdots + b_n\ S^n} \tag{7.1}$$

$$H_r(S) = \frac{\beta_0 + \beta_1 S + \beta_2 S^2 + \cdots + \beta_{n-1} S^{r-1}}{c_0 + c_1 S + c_2 S^2 + \cdots + c_n\ S^r} \tag{7.2}$$

$$a_1 c_0 + a_0 c_1 = b_1 \beta_0 + b_0 \beta_1 \tag{7.3}$$

$$\vdots$$

$$a_{n-1} c_r = b_n \beta_{r-1} \tag{7.4}$$

$$a_0 = \beta_0 \tag{7.5}$$

$$b_0 = c_0 \tag{7.6}$$

6 Conclusions

All proposed techniques have been tested on a wide range of examples. The results of the proposed method are compared with other popular methods presented in the literature and the results are excellent in terms of both speed and accuracy. Neural networks are the best in terms of accuracy and run time. Model order reduction using GA-SA hybrid solution proved to be a promising solution that can reduce complex high order models. The proposed methodology showed efficiency in terms of accuracy and preserving the basic properties of the original complex model over conventional mathematical methods increasing the opportunity for enhancing simulations and increasing the accuracy of modeling. There are new directions in machine learning: the chaos features can improve the efficiency of the genetic algorithm and SNNs have great potential for improving the accuracy and efficiency of conventional neural networks.

References

1. G. De Luca, G. Antonini, P. Benner, A parallel, adaptive multi-point model order reduction algorithm, in *22nd IEEE Conference on Electrical Performance of Electronic Packaging and Systems (EPEPS)*, 2013
2. J.E. Schutt-Ainé, P. Goh, Comparing fast convolution and model order reduction methods for S-parameter simulation, in *IEEE Electrical Design of Advanced Packaging & Systems Symposium (EDAPS)*, 2013
3. B. Vernay, A. Krust, T. Maehne, G. Schröpfer, F. Pêcheux, M.M. Louërat, A Novel Method of MEMS System-level Modeling via Multi-Domain Virtual Prototyping in SystemC-AMS, EDAA PhD Forum at DATE 2014
4. N. Kumar, K.J. Vinoy, S. Gopalakrishnan, Augmented Krylov model order reduction for finite element approximation of plane wave scattering problems, in *IEEE MTT-S International Microwave and RF Conference, IMaRC* 2013
5. H. Aridhi, M.H. Zaki, S. Tahar, Towards improving simulation of analog circuits using model order reduction, in *Design, Automation & Test in Europe Conference and Exhibition (DATE)*, 2012
6. A.C. Antoulas, D.C. Sorensen, S. Gugercin, A survey of model reduction methods for large-scale systems, Antoulas, Sorensen, et al., 2001
7. J.M. Wang, C.-C. Chu, Y. Qingjian, E.S. Kuh, On projection-based algorithms for model-order reduction of interconnects. IEEE Trans. Circuits Syst. I **49**(11) (2002)
8. Y.-L. Jiang, H.-B. Chen, Time domain model order reduction of general orthogonal polynomials for linear input-output systems. IEEE Trans. Autom. Control **57**(2), 330–343 (2012)
9. G. Parmar, M.K. Pandey, V. Kumar, System order reduction using GA for unit impulse input and a comparative study using ISE & IRE, in *International Conference on Advances in Computing, Communication and Control (ICAC3'09)*, 2009
10. E.S. Gopi, *Algorithm Collections for Digital Signal Processing Applications Using Matlab* (National Institute of Technology, Trichy, 2007)
11. Z.S. Abo-Hammour, O.M. Alsmadi, A.M. Al-Smadi, Frequency-based model order reduction via genetic algorithm approach, in *7th International Workshop on Systems, Signal Processing and their Applications (WOSSPA)*, 2011
12. M. Li, W. Du, L. Yuan, Feature selection of face recognition based on improved chaos genetic algorithm, in *Proceedings of the 3th International Symposium on Electronic Commerce and Security*, Guangzhou, 2010, pp. 74–78
13. S. Sen, S. Venkataramani, A. Raghunathan, Approximate computing for spiking neural networks, in *Design, Automation and Test in Europe (DATE)*, 2017
14. X. Lai, M. Zhang, An efficient ensemble of GA and PSO for real function optimization, in *2nd IEEE International Conference on Computer Science and Information Technology*, 2009

15. S.M. Odeh, Hybrid algorithm: fuzzy logic-genetic algorithm on traffic light intelligent system, in *FUZZ-IEEE*, 2015
16. M. Yaghini, *ANNSA: A Hybrid Artificial Neural Network/Simulated Annealing Algorithm for Optimal Control Problems* (Elsevier, 2012)
17. T. Sanguanchue, Hybrid algorithm for training feed-forward neural networks using PSO-information gain with back propagation algorithm, in *ECTICon*, 2012
18. X. Song, *Hybrid Algorithm of PSO and SA for Solving JSP* (IEEE, Shandong, 2008)
19. S. Agarwal, L.E. Pape, C.H. Dagli, A hybrid genetic algorithm and particle swarm optimization with type-2 fuzzy sets for generating systems of systems architectures. Procedia Comput. Sci. **36**, 57–64 (2014)
20. N. Ghaffari-Nasab, Hybrid simulated annealing based heuristic for solving the location-routing problem with fuzzy demands. Sci. Iran. **20**(3), 919–930 (2013)
21. S. Panda, J.S. Yadav, N.P. Patidar, C. Ardil, Evolutionary techniques for model order reduction of large scale linear systems. World Acad. Sci. Eng. Technol. **57**, 200–216 (2009)

Chapter 8
Conclusions

In this book, we presented the basic principles of applying so-called "machine learning" in model order reduction and highlighted the key benefits. One of the major research areas in CAD is circuit simulation. Circuit simulation is to use mathematical models to predict the behavior of an electronic circuit. A circuit is usually represented by a set of partial differential equations (PDEs) or ordinary differential equations (ODEs). So, the circuit simulation actually involves solving large-scale ODEs which sometimes takes several days or even weeks. Therefore, fast and accurate circuit simulation algorithms are needed to accelerate the simulation cycle. One way to speedup the simulation is to approximate the original system with an appropriately simplified system which captures the main properties of the original one. This method is called model order reduction (MOR), which reduces the complexity of the original large system and generates a reduced-order model (ROM) to represent the original one. This book presented novel MOR techniques based on machine learning such as genetic algorithm, neural network, simulated annealing, fuzzy logic, particle swarm optimization, and ant/bee colony. This book compared different approaches for model order reduction. In addition, hybrid solutions are presented. Finally, new directions in machine learning are discussed.

K. S. Mohamed, *Machine Learning for Model Order Reduction*,
https://doi.org/10.1007/978-3-319-75714-8_8

Index

K. S. Mohamed, *Machine Learning for Model Order Reduction*,
https://doi.org/10.1007/978-3-319-75714-8

The manufacturer's authorised representative in the EU is Springer Nature Customer Service Centre GmbH, Europaplatz 3, 69115 Heidelberg, Germany. If you have any concerns regarding our products, please contact ProductSafety@springernature.com

Printed and bound by CPI Group (UK) Ltd, Croydon, CR0 4YY
05/10/2026
02228152-0006